SpringerBriefs in Earth System Sciences

Series Editors

Gerrit Lohmann, Universität Bremen, Bremen, Germany

Lawrence A. Mysak, Department of Atmospheric and Oceanic Science, McGill University, Montreal, QC, Canada

Justus Notholt, Institute of Environmental Physics, University of Bremen, Bremen, Germany

Jorge Rabassa, Labaratorio de Geomorfología y Cuaternar, CADIC-CONICET, Ushuaia, Tierra de Fuego, Argentina

Vikram Unnithan, Department of Earth and Space Sciences, Jacobs University Bremen, Bremen, Germany

T0172159

SpringerBriefs in Earth System Sciences present concise summaries of cutting-edge research and practical applications. The series focuses on interdisciplinary research linking the lithosphere, atmosphere, biosphere, cryosphere, and hydrosphere building the system earth. It publishes peer-reviewed monographs under the editorial supervision of an international advisory board with the aim to publish 8 to 12 weeks after acceptance. Featuring compact volumes of 50 to 125 pages (approx. 20,000–70,000 words), the series covers a range of content from professional to academic such as:

- A timely reports of state-of-the art analytical techniques
- bridges between new research results
- snapshots of hot and/or emerging topics
- literature reviews
- in-depth case studies

Briefs are published as part of Springer's eBook collection, with millions of users worldwide. In addition, Briefs are available for individual print and electronic purchase.

Briefs are characterized by fast, global electronic dissemination, standard publishing contracts, easy-to-use manuscript preparation and formatting guidelines, and expedited production schedules.

Both solicited and unsolicited manuscripts are considered for publication in this series.

More information about this series at http://www.springer.com/series/10032

Manuel Enrique Pardo Echarte ·
Osvaldo Rodríguez Morán ·
Orelvis Delgado López

Non-seismic and Non-conventional Exploration Methods for Oil and Gas in Cuba

Springer

Manuel Enrique Pardo Echarte
Geology Scientific-Research Unit
Centro de Investigaciones del Petróleo
El Cerro, La Habana, Cuba

Osvaldo Rodríguez Morán
Geology Scientific-Research Unit
Centro de Investigaciones del Petróleo
El Cerro, La Habana, Cuba

Orelvis Delgado López
Geology Scientific-Research Unit
Centro de Investigaciones del Petróleo
El Cerro, La Habana, Cuba

ISSN 2191-589X ISSN 2191-5903 (electronic)
SpringerBriefs in Earth System Sciences
ISBN 978-3-030-15823-1 ISBN 978-3-030-15824-8 (eBook)
https://doi.org/10.1007/978-3-030-15824-8

Library of Congress Control Number: 2019935163

This Springer imprint is published by the registered company Springer Nature Switzerland AG
The registered company address is: Gewerbestrasse 11, 6330 Cham, Switzerland

Foreword

The successful exploration of hydrocarbon deposits culminates with the drilling of discovery wells of oil and gas reservoirs. The presence of a reservoir implies the formation and migration of hydrocarbons and their subsequent accumulation in a structural–stratigraphic geological trap. A deposit can be constituted by several reservoirs located at different depths, with different pressures and temperatures and with different types of hydrocarbons. They maintain in common to present a mixture of organic compounds with a great diversity of chemical compositions formed mostly by carbon and hydrogen (paraffinic, naphthenic, and aromatic series with the presence of oxygen, nitrogen, sulfur, and other impurities). All this diversity, together with the geological characteristics of each region, makes it more complex to detect them by conventional exploration methods, in addition to making this process more expensive. The optimum exploration technology to detect and map structures from the surface to depths of thousands of meters is seismic, in its different variants. However, the expected results are not always obtained for different reasons. For example, there is not always a marked acoustic impedance contrast between the different horizons of the geological section. Therefore, the use of non-seismic and non-conventional methods, that help to delimit and specify the results of the exploration, is justified.

There are many and diverse non-seismic and non-conventional methods. In Cuba, we have application experiences with positive results in several areas of the Cuban gas–oil regions. The authors of this monograph set out to show these positive experiences, using as evaluation criteria:

- The local gravimetric maximums, which reflect positive structures (by the more dense volcanic and carbonate uplifts), sometimes within the limits of certain values of the aeromagnetic field reduced to pole;
- Minimum values of the K/Th ratio, with local maximums of U (Ra) in its periphery;
- Positive residual geomorphic anomalies;
- Remote sensing anomalies;
- Anomalous indications by the **Redox Complex**.

Regarding the morphotectonic regionalization of the seas south of Cuba in shallow, transitional, and deep waters, applied to the region of the Batabanó, Ana María and Guacanayabo Gulfs, the Cayman Ridge, the Yucatan Basin and the Bartlett Trench, their results allowed to establish the areas of greatest prospective interest, in case of an active petroleum system.

This monograph compiles a large part of the knowledge and experience of the authors, applied to specific areas (Ciego de Ávila, Las Villas, and Sancti Spiritus, in central Cuba, and the Habana–Matanzas region in western Cuba), in order of complementing the results of seismic investigations, often not very resolutive. The results presented here will serve as reference material for specialists who wish to deepen the application of non-seismic and non-conventional methods for hydrocarbon exploration in areas with a complex geological structure.

The authors: Manuel E. Pardo Echarte, Geophysical Engineer, Doctor of Geological—Mineralogical Sciences, Osvaldo Rodríguez Morán, Geophysical Engineer, Doctor of Technical Sciences and Orelvis Delgado López, Geological Engineer, Master in Geophysics, have more than 15 years of work experience at the Petroleum Research Center, where they have spent most of their working lives in the Geology Basic Technical Scientific Unit.

El Cerro, La Habana, Cuba   Dr. Norma Rodríguez Martínez
                                        Production Basic Technical Scientific Unit
                                                    Petroleum Research Center

Preface

In various geological situations, seismic data provide little or no information about whether a trap is loaded with hydrocarbons or not. In other cases, when the acquisition is difficult and extremely expensive, or the quality of the information is poor due to geology or unfavorable surface conditions, it is the non-seismic exploration methods and, in particular, the unconventional methods of exploration, the only ones that can provide information about subtle stratigraphic traps. Besides, it is well documented that the generality of hydrocarbon accumulations has leaks or microseepage, which are predominantly vertical, as well as that they can be detected and mapped using various non-conventional and non-seismic methods of exploration. The benefits in the use of non-seismic and non-conventional exploration methods, integrated with geological data and conventional methods, translate into a better evaluation of prospects and exploration risk; such is the purpose that the book seeks.

In Cuba, there are two petroleum provinces, one north and another south. In the Northern Province, the main geological scenario where hydrocarbons are produced is related to a folded and thrusted belt, making it very difficult to identify, by seismic, the elements that make up the petroleum systems (source rocks, reservoirs, and seals). The South Province is characterized by tertiary basins deposited on volcanic rocks and ophiolites over thrusted on the North American continental paleomargin. In this province, the seismic is more resolutive than in the north, but there is the difficulty that the rock eval studies have not revealed the source rocks existing in that territory, which are demonstrated by the hydrocarbon shows in wells and on the surface. The majority of the hydrocarbon shows have been studied by the techniques of chromatography coupled to mass spectrometry (biomarkers) and grouped into genetic families that have subsequently been correlated with the source rocks identified in the Northern Petroleum Province through rock eval studies. From the previous data, the geographical limits and stratigraphic extensions of the petroleum systems in the Northern Petroleum Province have been defined. These spatial and temporal characteristics were extrapolated to all that territory, and in this way, the active systems were defined in the different regions of the north of the country. Thus, four stratigraphic intervals of source rocks (Callovian Middle

Jurassic, Oxfordian Upper Jurassic, Tithonian Upper Jurassic—Barremian Lower Cretaceous and Aptian Lower Cretaceous—Turonian Upper Cretaceous) and three oil genetic families (I, II, and III) were identified. In the case of the Southern Province, although there are several hydrocarbon shows that indicate the presence of active source rocks, the lack of knowledge of them does not allow the definition of petroleum systems. The biomarker data indicate that in the southern basins of Cuba, the petroleum systems identified in the Northern Province can coexist with other systems. These are associated with source rocks of the Upper Cretaceous or Tertiary age, according to the presence of Oleanano in oil shows obtained in the Ana María 1 well and in the sands of the keys of the Guacanayabo Gulf.

A geological task posed to the geological–geophysical processing and inter-pretation consisted in the mapping of possible new gaso-petroleum targets that will base the exploration in the Pina–Ceballos (northeast of the Central Basin) and Sancti Spíritus regions. In addition, an evaluation by recognition works of the **Redox Complex** of several of these possible new targets was envisaged. The mapping of the areas of interest was proposed based on the presence of a complex of indicator anomalies, mainly gravimetric, aeromagnetic, and airborne gamma spectrometric. To this end, the gravimetric (Bouguer reduction 2.3 t/m^3) and aeromagnetic field (reduced to pole) at 1:50,000 scale, the airborne gamma spec-trometry (channels K, Th, and U (Ra)) at scale 1:100,000 and the Digital Elevation Models 90×90 m and 30×30 m of the territory, were processed. The results indicate that the Pina oilfield anomalous complex is recognized, at least, in four other new localities, although with less areal extension; one of them is the Paraíso sector. Other deposits and prospects such as Brujo, Ceballos, and Pina Sur have anomalous complexes similar to Pina's, but incomplete in some of their attributes. The same happens for other established interest sectors. From the use of the **Redox Complex**, the presence of hydrocarbons in the depth was established in different sectors with anomalous indicator complexes, many of them coinciding with seismic structures. These sectors are, in Ciego de Ávila: Pina Sur, Pina Sur SO, Oeste de Ceballos 1, Oeste de Ceballos 2 and Pina Oeste Norte, and; in Sancti Spíritus, Gálata 2.

A version of the geo-structural mapping of the Habana–Matanzas (Block 7) region, based on the gravitational–magnetic data and, the mapping of sectors of gasopetroliferous interest linked to the conventional oil of the Placetas Tectonic-Stratigraphic Unit, is offered, based on the presence of a complex of indicator anomalies. The source materials were the same used for the previous region. Besides, a results' map of remote sensing for the search of perspective gasopetroliferous sectors in the region of Guanabo–Seboruco was used. The pro-cessing consisted in the regional-residual separation of the gravimetric and mor-phometric fields, the calculation of the first vertical derivative of the gravimetric and aeromagnetic fields, the inclination derivative of the reduced to pole magnetic field and of the ratio K/Th spectrometry channels. The indicator anomalous complex considers the following attributes: low-amplitude local gravimetric maxima; mini-mum of K/Th ratio and local maximums of U (Ra) at its periphery; local maximums of residual relief and remote sensing anomalies. As a result of the geo-structural mapping from the gravimagnetic data, a wide distribution of the Zaza Terrain (volcanic + ophiolites) was observed in the study region. The main structural depressions are concentrated along a latitudinal strip that covers the following locations (from east to west): Southwest of Matanzas Bay, Ceiba Mocha, Aguacate, Bainoa, Tapaste, Cuatro Caminos, Managua, and Santiago de Las Vegas. Based on estimates of the reduced to pole magnetic field, the depth at the top of a target located to the west of Bainoa, within the mentioned strip, was 1350–1450 m, which gives an idea of the basin's sedimentary thickness. The results of the integrated prospective cartography at the study region consider, in the first level of perspective, three localities (Boca de Jaruco, Jibacoa del Norte, and Este de Aguacate) where all the studied anomalies (attributes), with the exception of the morphometric ones, appear. In the second level of perspective, the localities that correspond to the combination of two types of different anomalies (11 localities) were considered.

Another geological task posed to the geological–geophysical processing and interpretation consisted in the mapping of possible new gaso-petroleum targets that will base the exploration at two regions in the western (Block 9) and central (Block 13) Cuba. The source materials and the processing were the same than the previous regions, with the exception of the remote sensing anomalies. As result, the mapping of sectors of oil–gas interest in western and central Cuba, related to the conventional oil of the Placetas Tectonic-Stratigraphic Unit and the Jurassic level, was based on the presence of a complex of indicator anomalies. It considers the following attributes: subtle local gravimetric maximum (in or near regional minimum); minimum of the K/Th ratio and local maximum of U (Ra) at its periphery, and; maximum of residual relief.

The last scenario of study includes the seas to the south of Cuba: shallow, transitional, and deep waters, characterized by the Batabanó, Ana María, and Guacanayabo gulfs, parts of the Yucatan Basin, the Cayman Ridge, and the Cayman Trench (on the edge of the Bartlett Fault). For this territory, with a low degree of seismic study, it was very important to count on other complementary non-seismic exploratory tools, like Digital Elevation Models, that allowed a preliminary assessment of the prospective areas. Then, the geological task posed to the morphometric processing and interpretation consists of the morphotectonic regionalization, with precision of the morpho-structural evidence of the so-called "Camagüey Trench" and, the establishment of possible sectors of oil and gas interest linked to the presence of geomorphic anomalies, presumably Indicators. For such purposes, the Digital Elevation Model 90 × 90 m was processed. As a result, the Morphotectonic Regionalization of the South Cuba marine territory was carried out, recognizing three types of regions: shallow waters (less than 100 m), transitional waters (greater than 100 and less than 3000 m), and deep waters (greater than 3000 m), from which could be characterized the different physiographic features already known. According to the interpretation, the existence of obvious signs of a trench and, therefore, of a paleo-subduction zone in southern Cuba is not recognized. Taking into account the range of amplitude of the geomorphic anomalies, if there is an active petroleum system, the areas of greatest prospective interest would correspond to the three gulfs (shallow waters) and, to the southwest and south of Batabanó Gulf and the northern limit of Cayman Ridge (transitional waters). The anomalies in deep water are vetoed by the huge water strain.

El Cerro, La Habana, Cuba

Manuel Enrique Pardo Echarte
Osvaldo Rodríguez Morán
Orelvis Delgado López

Acknowledgements

We thank our institution, Centro de Investigaciones del Petróleo (CUPET-Investigación), for allowing us to publish partial information concerning various research projects and, particularly, Figure 1.3, of its own scientific production.

We want to thank Juan Guillermo López Rivera, José Orlando López Quintero, Zulema Domínguez Sardiñas, Lourdes Jiménez de la Fuente, and Ramón Cruz Toledo for their technical support and supplied information.

We also want to thank for the partial or total revision of the manuscript and, for the correct observations to the same, of the following researchers: Dr. Evelio Linares Cala, Dr. Olga Castro Castiñeira, and Dr. Reinaldo Rojas Consuegra.

Contents

Abbreviations

AAE [Ea]	Average Activation Energy
AF [A]	Activation factor
AGS	Airborne gamma spectrometry
API	American Petroleum Institute, oil density measurements unit
CO_2	Carbon dioxide
CUPET	Cuba Petróleo
CVA [AVC]	Cretaceous volcanic arc
DEMs	Digital Elevation Models
Eh	Redox potential
Fm.	Geological formation
H_2S	Hydrogen sulfide
HCs	Hydrocarbons
HI [IH]	Hydrogen index
IP	Induced polarization
Ma	Millions of years
NCOS [FNPC]	Northern Cuban Oil Strip
OI [IO]	Oxygen index
OM	Organic matter
pH	Degree of acidity or basicity of an aqueous solution
Ro	Refractance of the vitrinite
RS	Remote sensing
S2	Generating potential of a source rock
SCPP	Southern Cuban Petroleum Province
Tmax	Maximum temperature
TOC [COT]	Total organic carbon
TSU [UTE]	Tectono-Stratigraphic Unit
UAC	Upward Analytical Continuation
VD	First vertical derivative

List of Figures

List of Tables

Chapter 1
Results of the Petroleum Systems Exploratory Method in Cuba

Orelvis Delgado López

Abstract In Cuba, there are two petroleum provinces, one north and other south. In the Northern Province, the main geological scenario where hydrocarbons are produced is related to a folded and thrusted belt, making it very difficult to identify, by seismic, the elements that make up the petroleum systems (source rocks, reservoirs and seals). The South Province is characterized by tertiary basins deposited on volcanic rocks and ophiolites over thrusted on the North American continental paleomargin. In this province, the seismic is more resolutive than in the north, but there is the difficulty that the rock eval studies have not revealed the source rocks (main and defining element of the petroleum systems) existing in that territory, which are demonstrated by the hydrocarbon shows in wells and on the surface. Although the known oil fields are in the Northern Petroleum Province, there are many oil and gas shows in isolated oil wells, in water wells and on the surface throughout the national territory. The majority of these shows have been studied by the techniques of chromatography coupled to mass spectrometry (biomarkers) and grouped into genetic families that have subsequently been correlated with the source rocks identified in the Northern Petroleum Province through rock eval studies. From the previous data, the geographical limits and stratigraphic extensions of the petroleum systems in the Northern Petroleum Province have been defined. These spatial and temporal characteristics were extrapolated to all that territory and, in this way, the active systems were defined in the different regions of the north of the country. Thus, four stratigraphic intervals of source rocks (Callovian Middle Jurassic, Oxfordian Upper Jurassic, Tithonian Upper Jurassic–Barremian Lower Cretaceous and Aptian Lower Cretaceous–Turonian Upper Cretaceous) and three oil genetic families (I, II and III) were identified. Families I and II were generated by the same level of source rock (Tithonian Upper Jurassic–Barremian Lower Cretaceous) product to facial variations. These variations of facies in the source rocks are something common in the mega basin of the Gulf of Mexico. The source rocks of the Aptian Lower Cretaceous–Turonian Upper Cretaceous generated Family III. In the case of the Southern

O. Delgado López (✉)
Centro de Investigaciones del Petróleo (Ceinpet), Churruca, no. 481, e/Vía Blanca y Washington, El Cerro CP 12000, La Habana, Cuba
e-mail: orelvis@ceinpet.cupet.cu

© The Author(s), under exclusive license to Springer Nature Switzerland AG 2019
M. E. Pardo Echarte et al., *Non-seismic and Non-conventional Exploration Methods for Oil and Gas in Cuba*, SpringerBriefs in Earth System Sciences, https://doi.org/10.1007/978-3-030-15824-8_1

Province, although there are several hydrocarbon shows that indicate the presence of active source rocks, the lack of knowledge of them does not allow the definition of petroleum systems. The biomarker data indicate that in the southern basins of Cuba the petroleum systems identified in the Northern Province can coexist (due to the presence of families II and III of Cuban crude oil) with other systems. These are associated with source rocks of the Upper Cretaceous or Tertiary age, according to the presence of Oleanano in oil shows obtained in the Ana María 1 well and in the sands of the keys of the Guacanayabo Gulf.

Keywords Hydrocarbon exploration · Petroleum systems · Source rocks · Oil families · Biomarkers · Cuba

1.1 Introduction

In Cuba, there are two petroleum provinces, one north and the other south, which present a very different geological scenario (Fig. 1.1). In the Northern Province, the main scenario where hydrocarbons are produced is related to a folded and thrusted belt, from the Jurassic to the Lower Eocene, that involves sediments of the North American continental paleomargin, Ophiolites and the Cretaceous Volcanic Arc (CVA-AVC). This geological complexity makes it very difficult to identify, by seismic, the elements that make up the petroleum systems (source rocks, reservoirs and seals) and the traps where the hydrocarbons are stored.

The South Province is characterized by tertiary basins deposited on volcanic rocks and ophiolites over thrusted on the North American continental paleomargin. In this province, the seismic is more decisive, but there is the difficulty that the rock eval studies have not revealed the source rocks (main and defining element of the petroleum systems) existing in that area, which are demonstrated by the hydrocarbon shows in wells and on the surface. That is why the possible stratigraphic levels of active source rocks in that province are predicted from biomarker data.

Although the known oil fields are in the Northern Province, in the entire national territory there are many oil and gas shows in isolated oil wells, in water wells and on the surface. These demonstrations have made possible the application of the exploratory method of Petroleum Systems, the central objective of this research, as well as the establishment of exploratory criteria to define the types of deposits to be found in the different areas of the northern part of the national territory. The results obtained have allowed the creation of geological-geochemical models, useful for the seismic interpretation in the Northern Province, with the consequent possibility of locating structural traps in this complex geological scenario.

1.1.1   Geographic Location and General Characteristics

Cuba is the largest island in the arch of the Greater Antilles and is surrounded by areas that are eminently oil-rich, such as the south of the United States of America, southeastern Mexico, and the Caribbean (Fig. 1.2).

The relief of the island is mostly flat, and the main watershed divides, in the east-west direction, the island into two parts, coinciding with the petroleum provinces defined in Cuba (Fig. 1.1). In the northern half of the island, there are elevations where the geological formations that constitute the elements of the petroleum systems identified in that territory, emerge. These hilly systems have facilitated the study of source rocks, reservoirs and seals. This research is developed throughout the national territory, including part of the shallow marine areas of the south of the country.

1.2   Theoretical Framework

In this section, a series of concepts on petroleum systems and information on Cuba's geology are presented, which will help to understand the results achieved.

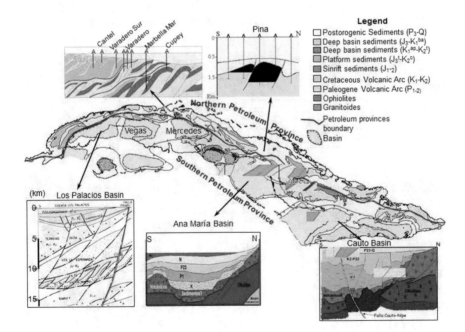

Fig. 1.1 Petroleum provinces and geological scenarios present in Cuba

1.2.1 Petroleum Systems Concepts

A petroleum system considers an active source rock and all accumulations of hydro-carbons genetically related to it. This includes all the elements and geological processes that are essential for the accumulation of oil and gas (Magoon and Dow 1994; Magoon and Beaumont 1999). According to these authors, the essential elements of a petroleum system are as follows: source rock, reservoirs and seals; while the processes that compose them are as follows: the burial of the source rock, the formation of the traps, processes of generation, migration, accumulation, and preservation of oil.

Source rock

It is a sedimentary rock with an organic matter content greater than 0.5% of its weight and that, under favorable conditions of temperature, is capable of generating and expelling hydrocarbons (Tissot and Welte 1984). The geochemical parameters that are used to characterize them are as follows:

(a) Source rock potential

It is determined by the amount of organic matter expressed in total organic carbon (TOC-COT, % total sample weight). It is considered a medium potential for TOC values between 0.5 and 1%, good for TOC values between 1 and 2%, very good

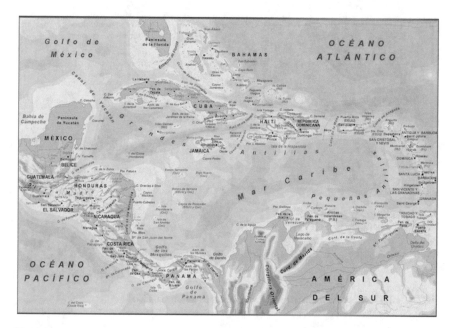

Fig. 1.2 Geographic location of Cuba in the context of the Caribbean–Gulf of Mexico region

for TOC values between 2 and 4%, and excellent for TOC values above 4%. The generating potential (S2) is also used, which means the quantity of hydrocarbons generated by one gram of source rock (mg HC/g rock). It is considered a medium potential for values of S2 between 2.5 and 5, good for values of S2 between 5 and 10, very good for values of S2 between 10 and 20, and excellent for values of S2 greater than 20.

(b) Type of organic matter

The type of organic matter that contains a source rock is an index of the quality of it. This is a very important parameter for characterizing a source rock because it determines the type of hydrocarbon expelled at the peak of maturation. Currently, six types of organic matter are recognized:

1. Type I: Organic matter of lacustrine origin. It has a large H_2 content in its molecular structure. It generates oil at the peak of thermal maturity.
2. Type II: Organic matter of marine origin. It has less H_2 content in its molecular structure than type I. It generates oil at the peak of thermal maturity.
3. Type IIS: Same origin as type II, but with a high content of S_2 in its molecular structure. It generates oil at the peak of thermal maturity.
4. Type III: Organic matter of terrestrial origin. It has little H_2 content in its molecular structure. It generates gas at the peak of thermal maturity.
5. Type II/III: Organic matter of terrestrial and marine origin (mixture). It generates oil and gas at the peak of thermal maturity.
6. Type IV: Inert organic matter, degraded or metamorphosed at the same moment of its deposition. It does not generate hydrocarbons.

Several analyzes are used to determine the type of organic matter, among which there found the kinetics of the kerogen, the hydrogen index (HI-IH), and the oxygen index (OI-IO) that are obtained by the pyrolysis rock eval.

(c) Thermal maturation

Thermal maturation is an index that expresses the temperature reached by a source rock and that allowed it to transform organic matter into hydrocarbons. The higher the level of maturity, the greater the volume of hydrocarbons generated and expelled, as well as the higher commercial quality of the same. The most commonly used geochemical parameters to determine maturation are vitrinite refractance (R_o, %) and maximum temperature (T_{max}, °C). The first is determined by optical methods and the second by pyrolysis rock eval. The ranges used for this parameter are as follows: immature, values of $R_o < 0.60$ and $T_{max} < 435$; early maturity, values of R_o between 0.60 and 0.65 and T_{max} between 435 and 445; maturity peak, values of R_o between 0.65 and 0.90 and T_{max} between 445 and 450; late maturity, values of Ro between 0.90 and 1.35 and T_{max} between 450 and 470; over maturity, values of $R_o > 1.35$ and $T_{max} > 470$.

Reservoir

It is any rock that provides the storage space of the hydrocarbons within the limits of the trap. For this it requires having an adequate porosity, which can be primary (depositional) or secondary (diagenetic or fracture), which allows it to store significant amounts of oil and/or gas. The reservoir must also have a good effective permeability that allows it to transfer and exchange fluids, that is, it can act as a conduit between the source rock and the trap (migration), displace the formation water that initially fills the trap and load it with migrated hydrocarbons and, subsequently, acting as the producing horizon.

Seal

It can be defined as a rock with little porosity and, above all, with a nearly zero connectivity between them (impermeable). In other words, any lithology that fulfills the condition of having a minimum displacement pressure greater than the fluctuating pressure of the hydrocarbons in the accumulation can act as a seal. In practice, the predominant lithologies are anhydrites, fine grained clastics and rocks with horizons enriched in organic matter. In general, there are three properties that determine the quality of a seal: (a) ductility, (b) thickness, and (c) uniformity.

(a) Ductility (plasticity): It is a property that varies with pressure and temperature as well as with lithology and refers to the ability to respond plastically (without fracturing) to tectonic efforts. In the thrust belts, where the deformations and fractures are very intense, this property is very important when studying the sealing horizons of the accumulations. Lithologies that have a very thin pore structure and a ductile matrix (clays) can maintain their sealing properties even under severe deformations.

(b) Thickness: Several centimeters of an ordinary clay are theoretically capable of trapping a large column of hydrocarbons. For example, a 10^{-4} mm clay can have a capillary pressure of 600 psi and theoretically be able to contain a vertical column of oil below it of 915 m. Unfortunately, it is very unlikely that in a thrusted area a horizon of a few centimeters thick may be continuous, not fractured and uniformly lithological over an accumulation. In this way, the increase in the thickness of a seal is proportional to its quality and to the probability of having a good sealing layer on a prospect.

(c) Uniformity: Uniformity is very important since it has been seen that small lithological variations can cause large variations of the capillary properties of the seals rocks.

It is important to note that there are two types of seals in a petroleum system:

(1) Regional seals: Are those that cover the migrated hydrocarbons and are characterized by having large extensions, significant thicknesses, and good lateral and lithological uniformity.

(2) Local seals: They are those that confine the accumulations and are much more complex than the regional ones. They are produced, mainly, by the activity of faults and must be carefully studied to determine their sealing characteristics in the three dimensions.

Burial of source rocks

It is the process of a petroleum system that allows a source rock to reach a state of thermal maturity to transform organic matter into hydrocarbons. To achieve this temperature, the source rock must be buried by an overload. This process can occur by continuous sedimentation, in quiet basins, or through tectonic stacking, in the thrust belts.

Trap Formation

Understand by trap to any geometric arrangement of rocks that allow significant accumulations of hydrocarbons. A trap includes the reservoir, which is the one that stores the fluids, and the seal, which prevents them from escaping. In nature, we can find different types of traps based on their genesis, and the most significant categories are as follows:

1. Structural

    (a) Associated with folds.
    (b) Associated with faults.
    (c) Combined.

2. Stratigraphic.

    (a) Primary or depositional.
    (b) Related to discontinuities.
    (c) Secondary or diagenetic.

3. Combined.

Structural traps

They are created by syn-postdepositional deformations of a stratum that give place to a structure allowing the accumulation of hydrocarbons. These are dominated by folds, faults or combinations of both.

Stratigraphic traps

They are those where the geometry of the trap and the seal–reservoir relationship was determined by a stratigraphic variation independent of structural deformations. They can be divided into three categories, based on their temporal relationship with sediment deposition: 1. Primary or depositional, 2. Related to discontinuities, 3. Secondary or diagenetic.

Combined traps

They are very discussed today. Some authors consider combined traps, those that owe their origin to both stratigraphic and structural processes and both are essential when evaluating them. However, at present, many researchers use the term to refer to traps that present stratigraphic and structural elements, being their main origin only one of the two processes.

Generation, migration, and accumulation of hydrocarbons

The processes of generation, migration, accumulation, and preservation of hydrocarbons are controlled by a group of factors:

1. Formation of sediments rich in organic matter (source rocks).
2. Preservation of organic matter rich in H_2.
4. Thermal maturation of organic matter.
5. Generation of a quantity of hydrocarbons that exceeds the retention capacity of the source rock.
6. The presence of porous, permeable strata, adjacent to the generating horizons.
7. Continuous compaction of the generating strata (overpressure).
8. Release of hydrocarbons in quantities that exceed the retention capacity of the transportation system.
9. Stratigraphic and structural interrelations that provide barriers to migration (traps).

Generation

It is the systematic transformation of organic matter into hydrocarbons, following specific diagenetic and metamorphic steps, product to the increase in temperature and time. In the literature, five zones of generation and alteration of hydrocarbons are described:

Zone 1. Corresponds to low subsurface temperatures. Gas can be generated from organic matter as a result of bacterial activity.

Zone 2. With the increase in the maturation of the kerogen, the amount of hydrocarbons generated increases, and the main products are dry and wet gas.

Zone 3. This is the main area of generation and expulsion of oil, and the main products are oil and gas.

Zone 4. Increasingly, lighter hydrocarbons (product of increased maturation) are generated from the kerogen and heavier hydrocarbons not expelled.

Zone 5. Intense organic metamorphism, complete carbonization of organic matter, dry gas product (methane).

Migration

It is the process by which the hydrocarbons, after being formed, are expelled from the source rock and find access routes to the trap where they can accumulate. Classically, the migration process can be divided into:

1. Primary migration or Expulsion.
2. Secondary migration.
3. Tertiary migration or Dismigration.

Primary migration or Expulsion

It is when a source rock saturated with hydrocarbons, product of the continuous thermal cracking of the kerogen, begins to expel it out of it due to the internal pressure caused by the generated fluids. This will depend on the type and enrichment of organic matter, as well as on its distribution within the source rock. In the literature, several mechanisms for primary migration are proposed:

1. Solution in water.
2. Diffusion through water.
3. Scattered drops.
4. Continuous oil phase.
5. Oil dissolved in gas.
6. Compaction pressure.
7. Internal pressure driven through induced fractures.

Secondary migration

It is the process by which the hydrocarbons expelled by a source rock are transported to the trap. Once the fluids reach the conductors (porous strata or faults), probably the most effective means for migration is a continuous phase of oil–gas. Its movement will be conditioned by vertical buoyant forces (given by the low density of oil compared to the formation water) that must exceed the capillary pressure of the conductive horizons. The speed of migration will depend basically on the viscosity and density of the fluid, the inclination, and characteristics of the conductive layers and the hydrodynamic activity of the geological environment. The total distance of the secondary migration of a hydrocarbon from the source rock to the trap can vary from a few hundred meters to a hundred kilometers.

Tertiary Migration or Dismigration

It refers to the escape of accumulated hydrocarbons to the surface (oil seep) through faults, discontinuities, or the seal itself, when the charge of the trap reaches a vertical column of fluids capable of having a buoyant pressure higher than the capillary pressure of the seal.

Accumulation

It occurs when the migrated hydrocarbons reach a trap and are stopped by a barrier that causes a progressive and relatively rapid (in the geological time) filling of these within the limits of the reservoir.

How to name a Petroleum System

To determine the presence of a petroleum system in a given area, we must find some hydrocarbon shows, no matter how small; this is proof of the existence of a petroleum system.

In the same way that geologists name geological formations, fossils, basins, etc., we must give a specific name to petroleum systems that distinguish them from others. The name of a petroleum system is composed of three parts: A–B (c), where:

A = name of the active source rock that generates the system, B = name of the reservoir that stores the largest number of reserves, c = symbol that expresses the level of certainty of the system. The level of certainty of a petroleum system can be: Proven, when there is a positive correlation oil-source rock and is expressed with the symbol (!); Hypothetical, when there is geochemical evidence and is expressed with the symbol (.) or; Speculative, when only geological and geophysical evidence is available and is expressed with the symbol (?).

Temporal and spatial aspects of a Petroleum System

A petroleum system is limited in time and space. Each system can be described in terms of its temporal and spatial elements and processes.

Temporary aspects of a petroleum system:

1. The Age: It is the time required for the processes of generation, migration, and accumulation of hydrocarbons to occur.
2. The Critical Moment: It is the geological time that best describes the previous process.
3. The Preservation Time: Starts immediately after the process of generation, migration, and accumulation and extends to the present.

Spatial aspects of petroleum systems:

1. Geographical extension: Curve that contours a focus of active source rock (oil kitchen zone) and all the shows, leaks, and accumulations of hydrocarbons originated by that focus. When the location of the oil kitchen zone is not known, it is called the geographical limit of the petroleum system.
2. Stratigraphic extension: It is the set of geological formations that enclose the essential elements of a petroleum system within its geographical extension.

1.2.2 Geological Premises

In general, the geological evolution of the Cuban territory can be divided into four major stages: rupture, drift, orogeny, and postorogeny.

Rupture stage (rift genesis)

The Gulf of Mexico was formed during two stages of rupture: 1. Upper Triassic–Lower Jurassic and 2. Middle Jurassic (Pindell and Barrett 1990; Sanchez Arango et al. 2001, 2007; Moretti et al. 2003). In this stage semigraben basins were developed and controlled the sedimentation. The depositional environments varied from continental to deltaic and neritic marine, because the primitive basins began to communicate with each other and, later, with the ocean (Sánchez Arango et al. 2001;

Cobiella Reguera and Olóriz 2009). In Cuba, these sequences are associated with the San Cayetano Formation (synrift). The aspects indicated, condition the deposition of organic matter of continental origin (type III) during this time (Middle Jurassic).

Drift stage

The continuation of the separation between North America and South America produced a thinning of the continental crust and the generation of oceanic crust, beginning the drift stage, which developed from the Upper Jurassic to the Turonian Upper Cretaceous. At the beginning of this stage, open spaces were created to the sea consequently with the sedimentation of shallow carbonates during the Kimmeridgian, which was succeeded by a sedimentation of the open sea during the Tithonian (Cobiella Reguera 2000, 2008; Pszczolkowski and Myczynski 2003; Sánchez Arango et al. 2001, 2007). From the above, it can be deduced that sediments deposited during the Kimmeridgian–Tithonian reflect a progressive change in the type of organic matter, from continental (type III) and/or algal (type I) in shallow marine environments, to type II in deep environments. Similarly, there is a continental or platform influence, through turbiditic currents, on the organic facies of deep sediments (Delgado López 2003).

In the Lower Cretaceous, the opening was completed in the southeastern Gulf of Mexico (Pindell 1994; Marton and Buffler 1999). The prevailing conditions from this moment, when there is a rapid subsidence in the basins of the Gulf of Mexico (Cobiella Reguera and Olóriz 2009), guarantee a predominance of marine organic matter (type II) in the Cretaceous sediments of the North American continental paleomargin. However, some organic component of continental origin (type III) could reach the basins by means of turbiditic currents, according to Pszczolkowski (1999), Cobiella Reguera et al. (2000) and Delgado López (2003).

To the objectives of this investigation, it is important the division that is made of the rocks of the North American continental paleomargin formed between the Upper Jurassic and the Upper Cretaceous (Sánchez Arango et al. 2001). In this interval, several Tectono-Stratigraphic Units (TSUs-UTEs) are identified, based on their geological history and their original paleogeographic position (Fig. 1.3). According to Sánchez Arango et al. (2001), the southernmost units have a greater continental contribution, while the northern units receive inflows from the platforms.

Orogenic stage

This stage develops from the Campanian Upper Cretaceous to the Middle Eocene. During this event, the collision between the Caribbean Volcanic Arc and the North American continental paleomargin gave rise to the structural relationship existing today between the different geological scenarios present in Cuba. This stage had a notable influence on several processes of the petroleum systems identified in the Northern Province:

- Fracturing of carbonate reservoirs of J_3–K_2 age, occurred.
- Regional seals were formed, associated with synorogenic sediments rich in clay of Upper Cretaceous and Paleogene age.

– The process of thermal maturation of the source rocks began, due to the increase in their burial by the tectonic stacking and sedimentary load.

In this stage, different tectonic mantles originated which are well recognized in the thrust belt, both in surface and in wells, from the west to central Cuba. From the oldest and deepest to the most shallow and young are (Fig. 1.1):

– Tectonic mantle of rocks from the Upper Jurassic to the Barremian Lower Cretaceous with synorogenic sequences, mostly of the Paleogene.
– Tectonic mantle of rocks of the Hauterivian–Turonian Cretaceous with synorogenic sequences, mostly of the Maastrichtian.
– Tectonic mantle of Ophiolites and CVA.

These tectonic mantles greatly influenced the spatial and temporal aspects of petroleum systems. This influence will be addressed in Sect. 4.3, related to the geographical limits, stratigraphic extension, and characterization of petroleum systems in Cuba.

Postorogenic stage

This stage was developed from the Upper Eocene to the Recent (Schlager et al. 1984; Tenreyro et al. 2001; Sanchez Arango et al. 2003) and was dominated by sedimentation of deep waters, with good stratification and clastic carbonated and

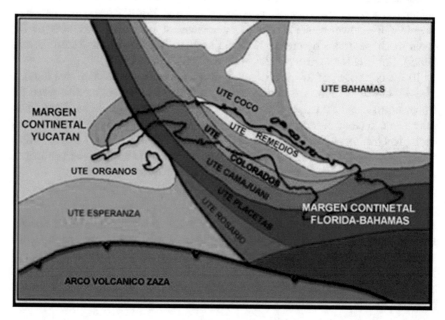

Fig. 1.3 Palinspastic reconstruction of the North American passive paleomargin before the orogeny

loamy composition. This stage also influenced the process of thermal maturation of the source rocks, although, unlike the previous stage, the influence on the burial was not as great as on the thermal history of the basin.

1.3   Materials and Methods

For the development of this work, a series of materials and methods were used that allowed obtaining data in the application of the exploratory method of petroleum systems.

1.3.1   Materials

Samples of drillings (cuttings) and cores from 80 wells were used, which added 1099 rock samples for studies of source rocks. These analyzes were complemented with 939 samples from 71 outcrops. For the characterization of the oil, samples were taken from oil fields, isolated producing wells, and dry wells. All shows were also sampled in geological mapping wells, water wells, and surface.

The drilling cores used were located in the Cuba Petróleo (CUPET) core store, while the drilling cuttings were obtained both, from the core storage (old wells), and from wells drilled in the last five years. The oil samples used were all sampled in the last years before carrying out this investigation.

1.3.2   Methods

The scientific tasks that are developed in this investigation describe the exploratory method of petroleum systems, which constitutes the second stage of the petroleum exploration process (Magoon and Dow 1994; Magoon and Beaumont 1999).

The starting point of this method is the study by biomarkers of all oil shows present in an area. Later, the oils are correlated to establish groups or families that indicate different source rocks. Then, the families of defined oils are correlated with the source rocks previously identified. The next step is the definition of the spatial and temporal aspects of the petroleum systems identified, which allows establishing the relationships between them.

To obtain the necessary data for the characterization of the source rocks and oils, the analytical techniques of pyrolysis rock eval and chromatography coupled to mass spectrometry were used, respectively. In addition, the results of the oil well temperature measurements were used to determine the geothermal gradients and calculate the oil and gas window depths in the different areas of Cuba.

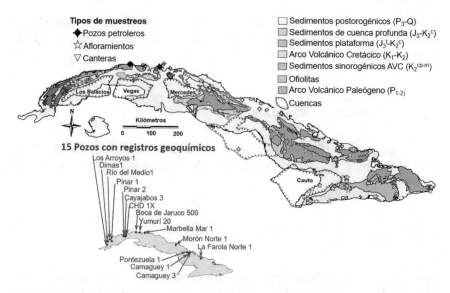

Fig. 1.4 Wells and outcrops used to collect sediment samples in studies of source rocks in Cuba. The outcrops of proven source rocks are located in the westernmost tip and in the central part of Cuba. In the east, Cenozoic sediments were sampled with negative results

1.4 Results

The results of the geochemical studies carried out in Cuba are presented, which include the following: the identification of the stratigraphic levels with properties of source rocks, the characterization of the oils and their genetic classification, as well as the oil-source rock correlations. With these data, the geographical limits, stratigraphic extensions, and characteristics of the petroleum systems in Cuba are established.

1.4.1 Cuban Source Rocks

Figure 1.4 shows a location map of the main samples of source rocks analyzed in Cuba. In most cases, it has been up to the operating company or the possibilities of CUPET to analyze samples in international laboratories. The database refers to samples taken from isolated outcrops or specific points in the section of a well.

Of the wells studied in western Cuba, they have a geochemical log: Marbella Mar 1, Puerto Escondido 2, Boca de Jaruco 501, and CHD-1X, which presents it incomplete. In recent field campaigns, detailed samplings have been carried out in different outcrops in order to compile geochemical records that characterize individual formations.

General characterization of the source rocks of Cuba

The main source rocks in all the TSUs of Cuba, as well as the largest oil reserves discovered, are related to the sediments of Kimmeridgian Upper Jurassic age that gradually transitions, without disagreement, to the Barremian Lower Cretaceous. There are also generating rocks in the Cretaceous (Aptian–Turonian). In addition, there are other intervals such as those of the Middle Jurassic and Oxfordian Jurassic of the San Cayetano Formation, with certain characteristics of source rocks.

Table 1.1 shows the most characteristic parameters, with values free of contamination risks and updated, for the different stratigraphic intervals and formations with conditions to be considered as source rocks in Cuba.

Kerogens

The kerogen studies are carried out fundamentally based on the results of the Rock Eval technique, particularly from the oxygen index (OI) versus the hydrogen index (HI), and the total organic carbon (TOC) versus generating potential (S2) ratios. Another type of analysis that is carried out is the determination of the kinetic parameters of the kerogen (bulk kinetic).

Figure 1.5 shows the great variety of kerogens present in the source rocks of Cuba (I, II, II/III, and III), even within the same formation, which is a consequence of the deposition environment of the sediments, as noted earlier in the geological premises. In the modified Van Krevelen chart, the predominance of type II organic matter is observed, indicating that the basin sediments of the westernmost part (San Cayetano, Órganos and Rosario TSUs) have a greater contribution of terrestrial plants in their kerogens, than those present in the central–eastern region; the latter having a significant influence of algal organic matter. On the other hand, there is a tendency in all the TSUs in the geological time to have a greater marine influence (Phytoplankton) in the younger sediments, which is in correspondence with the deepening that these basins suffered from the Tithonian. The presence of type IIS kerogens in the source rocks of the central and eastern regions of Cuba should be noted, as shown by the C/S ratios of some samples of the formations with characteristics of active source rocks and, because of the high sulfur content together with the low API gravities and maturity of the crude generated by them.

These variations of the organic facies in Cuban source rocks are confirmed in kinetic studies (bulk kinetic). In Fig. 1.6, some results of samples of the most important source rock interval in Cuba (J_3^{km}–K_1^{ba}) are shown, in which the variation of kerogens is observed.

In the figure, case A, organic facies very homogeneous, kerogen type I; case B, heterogeneous organic facies with low activation energy value, kerogen type IIS; case C and D, more homogenous organic facies, suggesting kerogen type II/III. In all cases, the average activation energy (AAE-Ea) and the activation factor (AF-A) are reported, as well as the pyrolysis data of the samples analyzed.

In summary, in the Cuban source rocks, seven organic facies have been identified, indistinctly in each one of them (Table 1.2).

Table 1.1 Stratigraphic intervals and geological formations with source rock properties in Cuba

Interval	Formation	TOC (%)	S2 (mg HC/g rock)	HI (mg HC/g TOC)	Organic matter type
Hauterivian–Turonian	Carmita, Polier, Alunado	0.37–14.9, 2.79 S: 54	0.72–27.63, 1.54 S: 38	25–751, 443 S: 39	II, II–III
Kimmeridgian–Barremian	Cifuentes, Jaguita, Artemisa, Ronda, Margarita, Sumidero	0.08–10.7, 1.18 S: 368	0.19–46.0, 8.75 S: 264	68–756, 343 S: 264	I, IIS y III
Oxfordian	Constancia, Francisco	0.26–3.64, 1.51 S: 34	0.56–10.44, 2.89 S: 34	110–383, 195 S: 34	I, II, II–III y III
Middle Jurassic	San Cayetano	1.21–5.57, 1.98 S: 26	0.17–1.92, 0.45 S: 26	117–431, 254 S: 26	III

The minimum, maximum, and average values of each parameter are reported, as well as the number of samples (S)

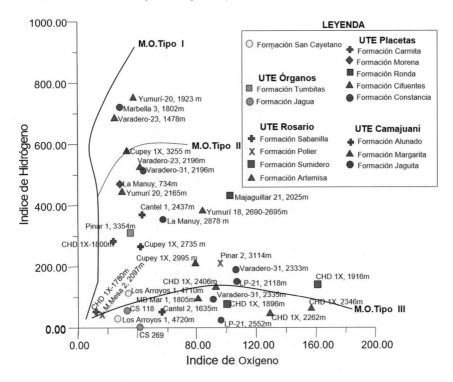

Fig. 1.5 Diagram Van Krevelen modified, showing the different types of organic matter present in the source rocks of Cuba

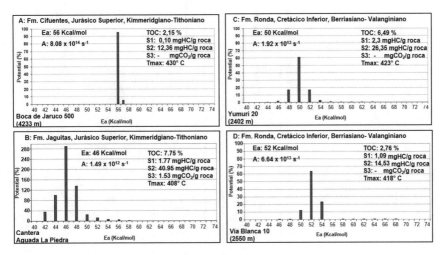

Fig. 1.6 Kinetic data of the source rocks of the Kimmeridgian Upper Jurassic–Barremian Lower Cretaceous interval, showing the presence of several types of organic facies

Table 1.2 Organic facies identified in the source rock intervals of Cuba from the rock eval and kerogen kinetics studies

Source rock interval	Organic facie A (inert organic matter)	Organic facie B (Type III organic matter, with Sulfur)	Organic facie C (Type III organic matter)	Organic facie D (Organic matter mix, Types II and III)	Organic facie E (Type II organic matter)	Organic facie F (Type II organic matter, with Sulfur)	Organic facie G (Type I organic matter)
K_1^{ap}–K_2^{t}	X		X	X	X		
J_3^{km}–K_1^{ba}	X		X	X	X	X	X
J_3^{ox}	X		X	X			
J_2	X	X					

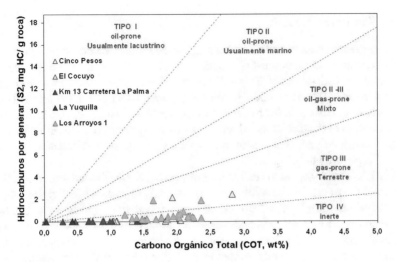

Fig. 1.7 Correlation graph TOC versus S2 for the samples corresponding to the San Cayetano Formation. The amount and type of organic matter present is shown

Potential of source rocks

The data that demonstrate the generating potential of the source rock intervals are presented below.

Middle Jurassic

At this age corresponds sedimentary units deposited during the phase of rift genesis, which have been studied in outcrops and wells in Cuba. They have been stratigraphically defined as: San Cayetano Formation, characterized by an alternation of sandstones, shales, and clays. Similar source rocks have been described in Mexico (Arellano 2001).

In order to evaluate the hydrocarbon generating potential for this age, data from 96 rock samples from nine outcrops and 27 samples from Los Arroyos 1 well, located at the westernmost tip of Cuba, were considered (Fig. 1.4).

Based on the compiled data, this unit presents variable values of TOC, which range between 0.21 and 7.86%, with an average value of 1.71%. The high variability of values allows us to suggest a vertical and lateral heterogeneity for this formation, which is in accordance with the geological data.

The correlation graph of TOC versus S2 was used to evaluate the generative quality of organic matter (Fig. 1.7). In general, most samples have a characteristic pattern of low potential to generate liquid hydrocarbons, even though levels with TOC values higher than 1.5% can be identified. That is, the low potential is not only linked to a low enrichment of organic matter, but also to the chemical composition of the organic matter deposited (inert, non-generative) or to the enrichment of type III or mixed organic material.

Oxfordian Upper Jurassic

It is important to remember that at this stage in the Gulf of Mexico began the change from terrigenous to carbonated sedimentation because the primitive basins began to communicate with the ocean. This will condition the existence of marine components in the kerogens (type II-III and II). Precisely, from the Oxfordian, the original basin begins to differentiate on the basis of the distance to the sources of continental input and extension of the platforms, which allows defining the TSUs. This differentiation of the basin conditions that source rocks of the same age have different amount and type of organic matter (Fig. 1.3).

The geological formations defined for this period are the Castellano, Jagua, Francisco and Constancia, whose lithology corresponds to rocks of fine grains, clayey, and carbonated (argillites, siltstones, clay schists, micrites, and fine intercalations of calcareous and quartzifers sandstones). To evaluate the hydrocarbon generating potential of this unit, data from 251 rock samples of outcrops from nine locations and 81 samples from eleven wells located in the western half of Cuba were considered (Fig. 1.4).

The existing data in this unit show variable values of TOC, ranging from 0.01% to 5.60%, with an average value of 0.96%. The high variability of values suggests a vertical and lateral heterogeneity of the organic facies of the Oxfordian source rocks, which is consistent with the geological data.

In the correlation graph of TOC versus S2 (Fig. 1.8), a clear differentiation was observed between the Oxfordian formations in the Placetas TSU with respect to the rest present in the basin. Samples from the Castellano, Jagua, and Francisco formations (western part) show a characteristic pattern of low potential to generate liquid hydrocarbons, even those with TOC values higher than 1.5%. The previous behavior indicates that the low potential is not only the product of a low enrichment in organic matter, but of the chemical composition of the same (inert, non-generative) and thermal maturity levels.

Kimmeridgian Upper Jurassic–Barremian Lower Cretaceous

The geological formations defined for this age range are: San Vicente, El Americano, Tumbadero, Tumbitas, Artemisa, Sumidero, Cifuentes, Ronda, Morena, Jaguitas, and Margarita. The lithology of these sediments represents a transgressive sequence from shallow waters (Kimmeridgian) to deep deposits (Tithonian–Barremian). During the Kimmeridgian, massive limestones or thick layers of light gray to black limestones and dolomites of shallow marine environments predominate, although at the end of this level there is a carbonated–clayey mudstone of black color. The Tithonian formations are composed of gray micrites, in thin layers with interspersed shales, and at the Berriasian level, the previous lithology is interspersed with lenses and layers of black chert.

From the above, it can be deduced that the sediments deposited during this period reflect a progressive change in the type of organic matter. From continental (type III) and/or algal (type I) in shallow environments, to marine (type II) in deep environ-

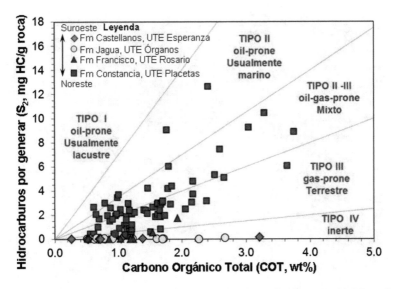

Fig. 1.8 Correlation graph TOC versus S2 for the samples corresponding to the Castellano, Jagua, Francisco, and Constancia formations. It shows the amount and type of organic matter present in the Oxfordian source rocks

ments. In the same way, there is a continental or platform influence (based on the latitude of the basin) through turbidites in the organic facies of the deep sediments.

To evaluate the hydrocarbon generating potential of this unit, data from 485 rock samples from 34 outcrops and 806 samples from 45 wells located in the western half of Cuba were considered (Fig. 1.4). The useful data in this range of source rocks show variable values of TOC, which range between 0.01 and 16.11%, with an average value of 1.05%. The high variability of values suggests a vertical and lateral heterogeneity in the organic facies, which is a consequence of the variations in the depositional environments of these sediments. For the analysis of the quality of the organic matter (OM) contained in the source rocks of this age, the samples were analyzed according to their original position in the basin (Fig. 1.3) to visualize, in this way, the regional trends. In the correlation TOC versus S2 shown in Fig. 1.9, the existence of several types of organic facies is clearly observed in all source rocks of this age range (Table 1.2).

The abundance of kerogens type IV (inert OM) in the TSU Rosario compared to the other units may be due to the fact that only in this TSU has taken place the development of continental margin magmatism, been recognized during the Tithonian Late Jurassic (represented by the Sábalo Formation). It could have metamorphosed part of the organic matter contained in the Artemisa Formation (J_3^{km-t}). The data that support this hypothesis are the high levels of thermal maturity recorded by the source rocks of the Rosario TSU when compared to those of the other units contained in the study (Fig. 1.10).

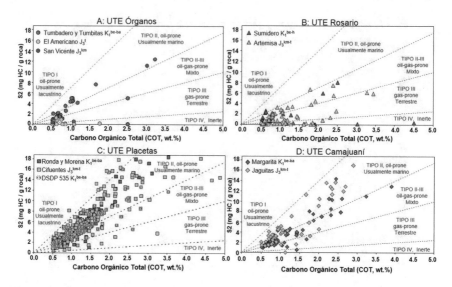

Fig. 1.9 Correlation graph TOC versus S2, showing the amount and type of organic matter present in the source rocks of the Kimmeridgian Upper Jurassic–Barremian Lower Cretaceous

Aptian Lower Cretaceous–Turonian Upper Cretaceous

The sediments deposited during this age interval correspond to the deepest depths of the basin (batial environments). However, in all the TSUs described in Cuba (Órganos, Rosario, Placetas and Camajuaní), the continental contribution through turbidites is evident. From the above, it can be deduced that the sediments deposited during this period will reflect a mixture of marine organic matter (type II), typical of the environment where they were deposited, with continental material (type III) provided by the turbidites.

The geological formations defined for this cycle are as follows: Pons, Polier, Santa Teresa, Carmita, Alunado, and Mata. The lithology of these formations is characterized by deep water sediments constituted by an alternation of micrites and silicites of dark colors. You can also find intercalations of shales, clays, and sandstones based on the position of the original basin.

To evaluate the hydrocarbon generating potential of this age range, data from 107 rock samples from 19 outcrops and 185 samples from 23 wells located in the western half of Cuba were considered (Fig. 1.4).

The data, free of contamination by migrated hydrocarbons, show variable values of TOC between 0.06 and 8.42%, with an average of 1.29%. The variability of values indicates a heterogeneity in the organic facies, which is a consequence of the depositional environments that these sediments had.

Figure 1.11 shows the correlation TOC versus S2 for this interval, indicating the existence of several types of organic facies (Table 1.2).

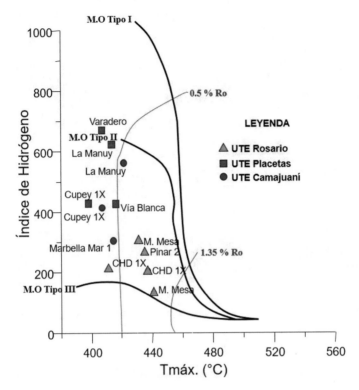

Fig. 1.10 Modified Van Krevelen diagram showing that the source rocks (J_3^{km}–K_1^{ba}) have higher thermal maturity in the Rosario TSU than in the other TSUs

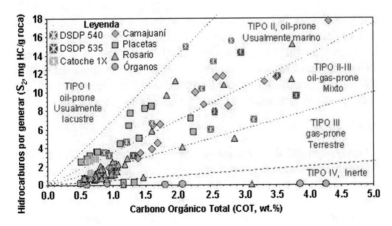

Fig. 1.11 Correlation graph TOC versus S2 for the samples corresponding to the source rocks of Aptian–Turonian age. The amount and type of organic matter is shown

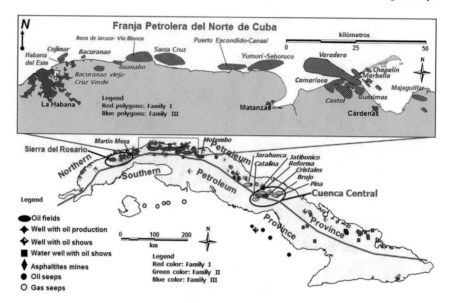

Fig. 1.12 Geographic location of oil deposits and shows in Cuba, showing the distribution of genetic families

1.4.2 Cuban Oil Families and Oil-Source Rocks Correlation

The Cuban hydrocarbons have been divided into three genetic families: Family I, Family II, and Family III. In Fig. 1.12, the geographic distribution of these families in Cuba is shown. Family I dominates the western part of Cuba, and it is related to the largest oil fields in the country. Family II is the most common in the central part, and it is associated with the vast majority of oil fields discovered in the provinces of Sancti Spíritus and Ciego de Ávila. Family III is the most abundant in the eastern part; however, it is present throughout the island, and it is associated with four small deposits of very good quality discovered in western and central Cuba.

Family I of Cuban oils

They are oils originated from marine organic matter, with or without sulfur, with a certain algal component, deposited in a somewhat anoxic siliciclastic carbonate marine environment ($\ll$0.1 ml oxygen/L of water).

On the basis of the molecular characteristics of this family, these oils were generated by the organic facies D, F, and G described above (Table 1.2). Considering that organic facies G and F are only present in Cuban source rocks of Kimmeridgian Upper Jurassic–Barremian Lower Cretaceous age, these are the sediments that generated Family I. The data of organic extracts obtained (Fig. 1.13 and Table 1.3) show that the great majority of these samples, obtained from the geological formations of the deep basin of age J_3^{km}–K_1^{ba} in the Northern Cuban Oil Strip (NCOS-FNPC)

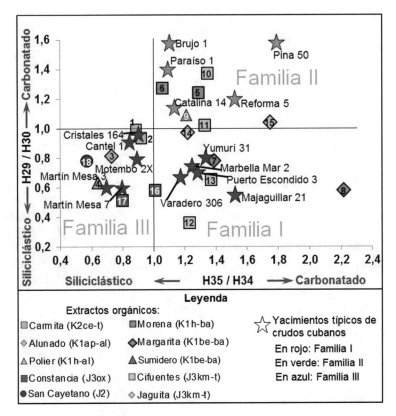

Fig. 1.13 Oil-source rock correlation for Cuban samples, according to biomarker relationships (Hopanos). For the identification of extract samples, use the code in Table 1.3

(Fig. 1.12), correlate with Family I. The only exception is the sample from well Puerto Escondido 2. Figure 1.12 shows that in the NCOS what is most abundant is Family I, which has a close molecular correlation with the source rocks described in that area.

Family II of Cuban oils

They are oils originating from marine organic matter, with sulfur, deposited in an anoxic carbonate marine environment (<0.1 ml oxygen/L of water).

Based on the molecular characteristics of this family, these oils were generated by organic facies E and F, described in the first part of this chapter. Due to that organic facies F is only present in Cuban source rocks of Kimmeridgian Upper Jurassic–Barremian Lower Cretaceous age, these are the sediments that generated Family II; this is due to lateral facial variations. Figure 1.12 shows a trend regarding the genetic type of hydrocarbons present in western and central Cuba, with a predominance of Family I, in the west, and Family II, in the central part. Data from organic extracts in

Table 1.3 Samples of organic extracts obtained from some Cuban source rocks

Code	Sample	Well or outcrop	Interval (m)	Formation	Location
1	940987018	Guásimas 41	1948–1952	Carmita (K_2^{ce-t})	NCOS
2	940990021	Boca de Jaruco 451	1251	Carmita (K_2^{ce-t})	NCOS
3	GQ160	Calienes Quarry	outcrop	Alunado (K_1^{ap-al})	Central Cuba (Villa Clara)
4	940989932	Chacón 2	1384–1389	Polier (K_1^{h-al})	Western Cuba (Pinar del Río)
5	AM-20	Loma Bonachea	outcrop	Morena (K_1^{h-ba})	Central Cuba (Villa Clara)
6	AM-22	Loma Bonachea	outcrop	Morena (K_1^{h-ba})	Central Cuba (Villa Clara)
7	AM-13	Aguada La Piedra	outcrop	Margarita (K_1^{be-ba})	Central Cuba (Villa Clara)
8	AM-14	Aguada La Piedra	outcrop	Margarita (K_1^{be-ba})	Central Cuba (Villa Clara)
9	EL-125-3-20	Las Lajas	outcrop	Sumidero (K_1^{be-ba})	Western Cuba (Pinar del Río)
10	AM-17	Loma Sin Nombre	outcrop	Cifuentes (J_3^{km-t})	Central Cuba (Villa Clara)
11	940990456	Puerto Escondido 2	4002–4005	Cifuentes (J_3^{km-t})	NCOS
12	940990532	Varadero 201	2600–2604	Cifuentes (J_3^{km-t})	NCOS
13	940990155	Litoral Pedraplén Mar 1	2105–2110	Cifuentes (J_3^{km-t})	NCOS
14	GQ180	Loma Las Azores	outcrop	Jaguita (J_3^{km-t})	Central Cuba (Villa Clara)
15	GQ190B	Aguada La Piedra	outcrop	Jaguita (J_3^{km-t})	Central Cuba (Villa Clara)
16	940900952	Varadero 201	2829–2834	Constancia (J_3^{ox})	NCOS
17	940990299	Litoral Pedraplén 21	2393–2396	Constancia (J_3^{ox})	NCOS
18	E-053	Cinco Pesos	outcrop	San Cayetano (J_2)	Western Cuba (Pinar del Río)

Cuba (Fig. 1.13 and Table 1.3) indicate that the vast majority of the extracts obtained from the J_3^{km}–K_1^{ba} source rocks of Central Cuba correlate with Family II of Cuban crudes. In works carried out in the mega basin of the Gulf of Mexico (Rocha Mello and Trindade 1996; Guzmán Vega et al. 2001), facial variations are also identified in the source rocks of age J_3^{km}–K_1^{ba} as well as different genetic families of associated hydrocarbons to them.

Family III of Cuban oils

They are oils originated from marine organic matter mixed with continental, deposited in a suboxic carbonate-siliciclastic marine environment (<0.2 ml oxygen/L of water).

On the basis of the molecular characteristics of this family, these oils were generated by the organic facies C, D, and E described in Sect. 4.1.1. These three organic facies are present in two intervals of Cuban source rocks (J_3^{ox} and K_1^{ap}–K_2^t) which could generate hydrocarbons from Family III. However, Fig. 1.13 shows that all organic extracts available from the geological formations of age K_1^{ap}–K_2^t correlate with Family III, independently of the TSU to which they belong. This aspect indicates the regional behavior in terms of the age of the Cuban source rocks and the genetic type of oil (family) that these generate. Note in the graph, the close correlation between the extracts of the Carmita Formation (Fm.) with the crude oils from the Cantel and Cristales (central region of Cuba) oil fields, while the oils from the Martín Mesa Field correlate more with the extract from the Polier Fm. (western region of Cuba).

1.4.3   Geographic Limits, Stratigraphic Extension and Characterization of Petroleum Systems in Cuba

In Cuba, three genetic types of hydrocarbons have been found to date (families I, II, and III of Cuban crudes, Figs. 1.12 and 1.13). The presence of three genetic types of hydrocarbons would indicate the existence of three petroleum systems, one related to each family (Magoon and Dow 1994)l however, the oil-source rock correlations established in Cuba (Fig. 1.13) indicate that several of the recognized source rocks can generate oils of Family I. These source rocks belong to the formations of the Veloz Group of the Placetas TSU (Cifuentes, Ronda and Morena), the Jaguitas and Margarita formations of the Camajuaní TSU, as well as the Artemisa and Sumidero formations of the Rosario TSU. These data point to the presence of at least three different petroleum systems that generated oils of the Family I, one related to the Placetas TSU, another with the Camajuaní TSU, and a third associated with the Rosario TSU.

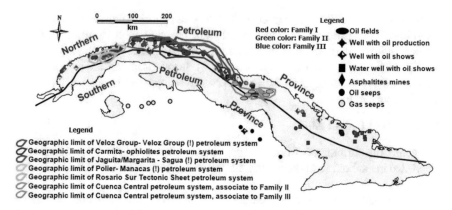

Fig. 1.14 Genetic types of oil and geographical limits of petroleum systems present in the Northern Petroleum Province. The area where these systems are best studied and characterized is in the Habana–Cárdenas region

Figure 1.14 shows the geographical limits of the petroleum systems identified in Cuba. Each one of them is presented briefly, in order of importance. In addition, the commercial qualities of the oils that produce each of the plays associated with these petroleum systems are mentioned, as well as the stratigraphic extensions of the same.

1.4.3.1 Veloz Group–Veloz Group (!) Petroleum System

With this system, the vast majority of the oil fields present in the NCOS are related, as well as the wells San Antón 1X, Martí 2, and Bolaños 1 and the oil seeps San Felipe, Angelita and Santa Gertrudis. The exceptions are the Cantel, Motembo, Cupey, and Chapelín oil fields; and the lower mantles of Majaguillar and Litoral Pedraplén-Marbella Mar; as well as the Martí 5 well (Figs. 1.12 and 1.14).

Oils

The oils of this petroleum system belong to the Family I of Cuban crude oil. From the point of view of their commercial quality, they are mostly heavy and highly sulfurous, caused by the presence of organic matter of the type IIS in the source rock that generated them and the low level of maturation reached by it. These hydrocarbons are prone to undergo fractionation processes which causes that coexists different qualities in different reservoirs of the same oil field; the tendency is that the quality decreases with an increase in depth. This process is well studied in the Vía Blanca Oil Field (Fig. 1.15). Note in this figure that, in the chromatogram of the upper reservoir, only light fractions are detected (NC8-NC12), while in the lower one, heavier compounds (NC16-NC32) are detected. To a lesser extent they can be affected by biodegradation.

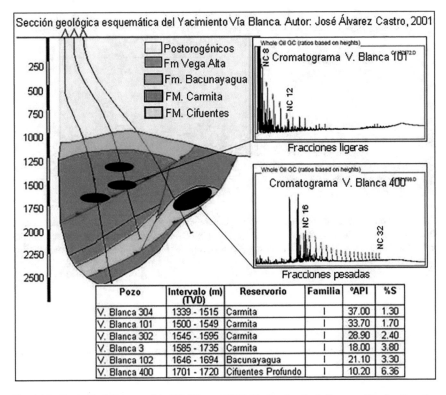

Fig. 1.15 Geological diagram of the Vía Blanca Oil Field, showing the influence of the fractionation process on the variation of oil quality

Source rock

With the existing data, to date, it can be stated that the source rock of this system is the Veloz Group (J_3^{Km}–K_1^{ba}) that groups the Cifuentes (J_3^{km-t}), Ronda (K_1^{be-v}), and Morena (K_1^{h-ba}) formations (Fig. 1.13).

Reservoirs

The main reservoirs are the sediments of the North American continental paleomargin of the Placetas TSU and the synorogenic ones related to it. In order of importance are Veloz Group (J_3^{Km}–K_1^{ba}), Carmita (K_2^{c-t}), Bacunayagua (K_2^{cp}), Amaro (K_2^{cp-m}), and Vega Alta (P_1–P_2^2) formations. Another reservoir present in this system are the Ophiolites, to which some wells of the Camarioca, Varadero Sur, Guásimas, and Boca de Jaruco oil fields are associated as well as the Cruz Verde and Bacuranao oil fields. Oils of this petroleum system have also been found in reservoirs of Middle Eocene age that cover the thrust belt, which demonstrates the great potential for loading it. In Fig. 1.16, with red dotted lines, the stratigraphic extension of this system is shown.

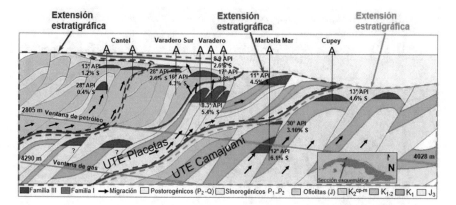

Fig. 1.16 Petroleum-geological scheme of the Cantel–Cupey area, showing how the tectonics of thrusts cause the sediments of the North American continental paleomargin to act as source rock and reservoir at the same time. In addition, the different tectonic mantles behave as independent petroleum systems

Seals

The regional seal is the Vega Alta Formation (P_1–P_2^2). Other seals related to this system have a local character and are the Ophiolites and the Vía Blanca Formation (K_2^{cp-m}).

Traps

The main traps related to this petroleum system are combined and refer to the frontal zone of the folded and thrusted mantles of the North American continental paleomargin. A second type of trap present corresponds to fractured areas within the mantles of the ophiolites.

Veloz Group Play

The largest volumes of oil resulting from the secondary migration in this system are stored in the Veloz Group, that acts as both source rock and reservoir. As a source rock, it operates at depths between 3000 and 4000 m in the Habana–Corralillo region (Table 1.4) and, as a reservoir, in the frontal zone of the scales (Fig. 1.16).

The migration and loading of this play occurs mostly subhorizontally through the tectonic scales of the Veloz Group, from the oil kitchen zone to the traps on the thrust fronts (Fig. 1.16). The crude oil that it stores is mostly heavy, due to the low thermal maturity reached by the source rock in the Habana–Cardenas region.

Carmita Play

This play stores smaller reserve volumes than the previous one, and oil accumulation occurs mainly as a consequence of tertiary migration. The crude oil that it stores has better quality than the play described above, sometimes reaching to be light crudes.

Table 1.4 Wells with studies of geothermal gradients in the Habana-Cárdenas region

Well	X coord.	Y coord.	Geothermal gradient	Deep for 80 °C (biodegradation)	Deep for 110 °C (Oil window)	Deep for 165 °C (Gas window)
Via Blanca	390066	372279	22.10	2488.69	3846.15	5882.35
Boca de Jaruco	394599	372493	22.20	2477.48	3828.83	5855.86
Yumurí	434250	369150	23.10	2380.95	3679.65	5627.71
Camarioca	463859	362101	30.30	1815.18	2805.28	4290.43
Guásimas	472510	366540	26.60	2067.67	3195.49	4887.22
Cupey 1X	473419	361845	24.80	2217.74	3427.42	5241.94

This behavior is due to the fact that its hydrocarbons are the light fractions of oil accumulated in the deepest play (Veloz Group) as a consequence of the oil fractionation process (Fig. 1.15). It is common for these oils to be affected by moderate biodegradation (e.g., Varadero Sur 13 well), because they are shallower and have a lower relative sulfur content than those of the previous play. Even though they are affected by biodegradation, they have better quality than those of the play Veloz Group for being the light products of the fractionation process.

The rest of the reservoirs known today that are related to this petroleum system (Bacunayagua, Amaro, Vega Alta, Ophiolites) store crude with commercial qualities determined by processes similar to the Carmita Play. All of the above indicates the existence of two main directions of entrapment. One linked to the folded and thrusted structures of the Veloz Group and another to the Middle Cretaceous (Carmita, Bacunayagua and Amaro formations) and Ophiolites. They highlight the enormous wealth of the first, and the lower depth and higher quality of the crude trapped in the second. The integration of the elements and processes of this petroleum system are presented in Fig. 1.17.

1.4.3.2 Carmita–Ophiolites (!) Petroleum System

With this petroleum system, the Cantel and Motembo oil fields are related; as well as the wells Martí 5 and Camarioca 6 and the oil seeps San Miguel de los Baños and Arroyo Biajaca. The Carmita–Ophiolites (!) system is developed in areas where the Carmita Formation has great thicknesses, resulting in repetitions of scales from the tectonic mantle of the Middle Cretaceous of the Placetas TSU. This assertion is made based on the results of wells recently drilled (Cantel 2000 R and Angelina 100) that cut large thicknesses of the Carmita Formation (Pérez et al. 2007; López Corzo et al. 2008).

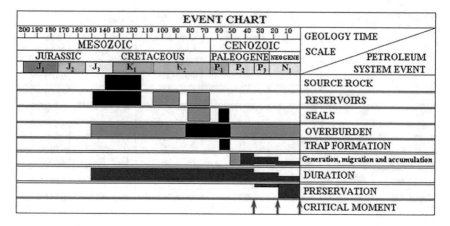

Fig. 1.17 Diagram of events of the Veloz Group - Veloz Group (!) Petroleum System

Oils

The oils of this petroleum system belong to the Family III of Cuban crude oil. Genetically they are light crude and not very sulfurous due to the source rock that generates them (Carmita Formation). They originated in a sub-oxic and clayey-carbonated environment, which guarantees that the kerogen contained in the source rock is free of sulfur. Because they are originally sweet oils, the hydrocarbons in this system are very attractive for oil-degrading bacteria, so biodegradation is the main secondary process that affects them. In this way, in the same oil fields, we can find oils from light to heavy; the tendency is to increase commercial quality with increasing depth. A good example of this process is in the Cantel Oil Field, in which the commercial quality of the oil varies by well despite having a common origin and identical maturity levels (Fig. 1.18). Note that the shallow reservoirs of this oil field (affected by biodegradation) store heavy crudes, while in the deepest ones the hydrocarbon reaches the category of medium. However, even in the most biodegradable crude, the sulfur content is low, oscillating, mostly, between the low sulfur (<0.5%) and mildly sulfurous (0.5–1.5%) categories.

Source rock

With the existing data, to date, it can be stated that the source rock of this system is the Carmita Formation (Fig. 1.13) (Delgado López 2003).

Reservoirs

The main reservoirs of this system are related to the Ophiolites, where the largest reserves are found. A second reservoir refers to the Middle Cretaceous sediments of the Placetas TSU, highlighting the Carmita Formation which acts as a source rock in depth and as a reservoir in the frontal zone of the scales from the tectonic mantle of the Middle Cretaceous of the Placetas TSU (Figs. 1.16 and 1.18). In Fig. 1.16, with blue dotted lines, the stratigraphic extension of this system is shown.

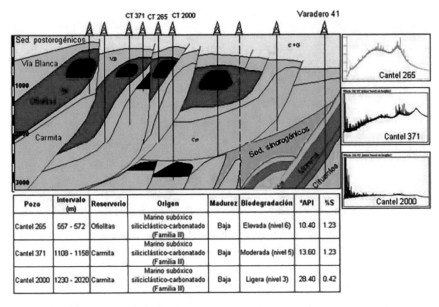

Pozo	Intervalo (m)	Reservorio	Origen	Madurez	Biodegradación	°API	%S
Cantel 265	567 - 572	Ofioltas	Marino subóxico siliciclástico-carbonatado (Familia III)	Baja	Elevada (nivel 6)	10.40	1.23
Cantel 371	1108 - 1158	Carmita	Marino subóxico siliciclástico-carbonatado (Familia III)	Baja	Moderada (nivel 5)	13.60	1.23
Cantel 2000	1230 - 2020	Carmita	Marino subóxico siliciclástico-carbonatado (Familia III)	Baja	Ligera (nivel 3)	28.40	0.42

Fig. 1.18 Geological diagram of the Cantel Oil Field, according to Sosa Meizoso (2001), showing the influence of the biodegradation process on the variation of oil quality

Seals

Today, a seal of a regional nature is not well defined for this petroleum system. It is considered that it is the Vía Blanca Formation that acts as a seal of the reservoirs of the Middle Cretaceous of the Placetas TSU. It could be that precisely the absence of a good regional seal in this system is the reason why the largest reserves are in the Ophiolites. A seal of local character are the ophiolites in the Cantel and Motembo oil fields.

Traps

The main traps associated with this system are fractured zones within the Ophiolites, in the mantles thrusted on the North American continental paleomargin (Figs. 1.16 and 1.18). Another type of traps present is of combined structural type and are related to the frontal zones of the thrusted mantles with sediments of the Middle Cretaceous of the Placetas TSU (Figs. 1.16 and 1.18).

Plays

Unlike the previous petroleum system, in this system there are few plays; so far only three are known. Next, they are characterized taking into account the hydrocarbon properties they store.

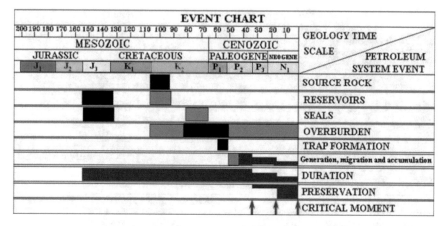

Fig. 1.19 Diagram of events of the Carmita–Ofiolitas (!) Petroleum System

Ophiolites Play

This is the play that produces the largest volumes of oil in this petroleum system (about 60% of the production accumulated in the Cantel Oil Field). This situation must be given because the ophiolites are rocks of impermeable nature which prevents the leakage of the light and slightly viscous hydrocarbons trapped in the fractured zones inside the thrusted ophiolites, out of sequence on the Placetas TSU. Most of the crudes in this play are biodegraded by being at shallow depths and temperatures below 80 °C.

Carmita–Vía Blanca Play

In this play, the traps constituted by the Carmita (reservoir) and Vía Blanca (seal) formations store oil of Family III; this being the great difference with the play of the Veloz Group–Veloz Group (!) petroleum system that has identical reservoirs and seal. This play accumulates little volume of hydrocarbons since the properties of the seal are not suitable to retain non-viscous and light-medium crudes as those associated with it. Generally, the hydrocarbons in this play have higher quality than those described above because they are deeper and protected from biodegradation (Fig. 1.18).

Bacunayagua–Vía Blanca Play

This play is known in the Camarioca 6 well and is linked to the Family III of Cuban crudes, not to the I as the rest of the Camarioca Oil Field. The crude oils stored in it are of, mostly, medium and moderately sulfurous qualities. Its main exploratory risk is the quality of the seal as well as the previous play. Figure 1.19 shows the integration of all the elements and processes of this system.

1.4.3.3   Jaguita/Margarita–Sagua (!) Petroleum System

This system is associated with the Camajuaní TSU and, the Cupey and Chapelín oil fields belong to it as well as the lower mantles of Majaguillar and Litoral Pedraplén-Marbella Mar. At present, it is little studied because the plays of the aforementioned TSU have been little exploited.

Oils

The oils linked to this petroleum system classify in the Family I of Cuban crudes. From the point of view of their commercial quality, they are mostly heavy and sulfurous, caused by the presence of organic matter of the type IIS in the source rock that generated them and the low level of maturation reached by it. These hydrocarbons, like those of the Veloz Group–Veloz Group (!) system, are prone to undergo fractionation processes, which leads to the coexistence of different qualities of oil in different reservoirs at the same oil field. The tendency is that the quality decreases with an increase in depth. An example of this process is found in the oils of the Marbella Mar 2 well, where the trapped at 2600 m in the Sagua Formation have better quality when compared to those found in the Jaguita Formation at 3180 m (Fig. 1.16). To a lesser extent, they can be affected by biodegradation.

Source rock

On the basis of the existing oil-source rock correlations (Fig. 1.13), the Jaguita and Margarita formations possess the sediments that generate this system. These units cover the stratigraphic Tithonian–Barremian interval of the Camajuaní TSU, constituting a coeval of the Veloz Group of the Placetas TSU. Many of the source rocks of the Gulf of Mexico are of this age and are grouped in the so-called Tithonian Generating Subsystem (González and Holguín 2001; Santamaría and Horsfield 2001).

Reservoirs

The reservoirs from where the greatest productions have been reached to date are the breccias of the Sagua Formation, from the Lower Eocene. However, it is possible that with the future development of the plays of this system greater productions of the reservoirs of the Neocomian Upper Jurassic interval will reach based on their quality (Valladares et al. 1996).

Seals

The regional seal of this system is the Vega Formation of Upper Eocene age, while other local seals (Valladares et al. 1996) constitute clayey facies within the Tithonian Upper Jurassic–Cenomanian Upper Cretaceous reservoirs.

The diagram of events for this petroleum system is presented in Fig. 1.20.

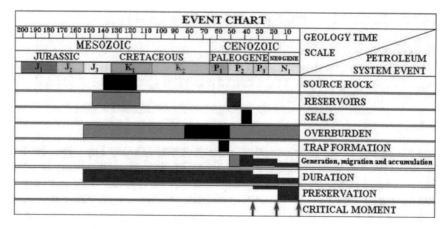

Fig. 1.20 Diagram of events of the Jaguita/Margarita-Sagua (!) Petroleum System

1.4.3.4 Polier–Manacas (!) Petroleum System

This system is linked to the Sierra del Rosario TSU, and it is associated with the Martín Mesa Oil Field, and all the oil shows of the Family III present in the Sierra del Rosario region. In Fig. 1.14, the geographic limit of this system is presented from the area of Burén (west) to Bejucal (east). In this petroleum system, there are particularities in several of its processes that differentiate it significantly from the rest of the systems present in Cuba.

Oils

The oils of this petroleum system belong to the Family III of Cuban crude oil. Genetically they are light crude and not very sulfurous because the source rock that generates them (Polier Formation) originated in a sub-oxic and clayey-carbonated environment that guarantees that the kerogen contained in the source rock is free of sulfur. Because they are sweet, they are usually affected by biodegradation, which causes oil quality to vary from heavy (18.7° API, in Martín Mesa 20 A well) to light (32.5° API, in Martín Mesa 7 well).

Source rock

The source rock of this system is the Polier Formation (K_1^{v-al}), as indicated by the oil-source rock correlation (Fig. 1.13). These sediments had sedimentation conditions similar to the Carmita Formation (K_2^{c-t}) of the Carmita–Ophiolites (!) system, described above.

Reservoirs

Oil productions can be indicated in the Manacas Fm. (Paleocene) or top of the mantles of the Polier Formation (K_1^{v-al}). Some significant productions of gas are associated

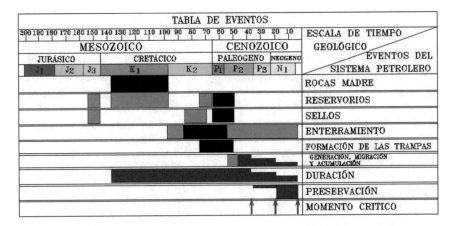

Fig. 1.21 Diagram of events of the Polier–Manacas (!) Petroleum System

to the Cacarajícara Fm. (K-Pg) as well as, although to a lesser extent, productions of the Vía Blanca Formation of the Campanian–Maastrichtian Upper Cretaceous are known.

Seals

The regional seal of the Polier–Manacas (!) petroleum system is constituted by the lower member of the Manacas Formation. The existence of local seals, so far, can only be related to the Ophiolites.

Traps

For this system, they are structural, associated with folding, and are not related to the thrust fronts, but to the middle parts of the mantle. This aspect is due to the fact that the tectonic mantle of Rosario Norte (Pszczółkowski 1999), where this system is found, is an out of sequence thrust (Iturralde Vinent 1996) and, its front is overturned and sunken to the north below thrusted mantles of the Ophiolites and the CVA. Particularly in the area of the Martín Mesa Oil Field, the complex and imbricated nature of tectonics make that the deposits are small and associated with the domes, where different formations act as a reservoir.

Figure 1.21 shows the diagram of events of the Polier–Manacas (!) petroleum system based on the integration of its elements and processes under the aspect of geological time.

1.4.3.5 Petroleum System Associated with the Rosario Sur Tectonic Mantle

It is convenient to point out that in the Sierra del Rosario TSU, there are no known oil fields related to families I and II. However, the degree of outcropping of the source

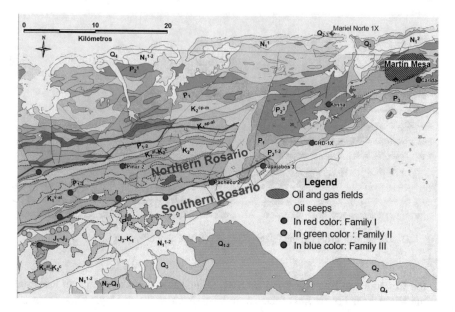

Fig. 1.22 Geological map at scale 1: 250,000 of the area occupied by the Sierra del Rosario TSU

rocks and the abundance of superficial oil shows allow to identify a petroleum system linked to the interval of source rocks of Kimmeridgian Upper Jurassic–Barremian Lower Cretaceous age.

Figure 1.22 shows the geological map at scale 1: 250,000 of the area occupied by the Sierra del Rosario TSU. In it, the division proposed by Pszczółkowski (1999), (Rosario Norte and Rosario Sur) is clearly observed based on the age of the Mesozoic rocks present in each tectonic mantle. In the Rosario Sur mantle, the rocks of the Upper Jurassic (Artemisa Fm., J_3^{km-t}) and Lower Cretaceous (Sumidero Fm., K_1^{be-v}) predominate, while in the Rosario Norte mantle, the Valanginian–Campanian Upper Cretaceous sediments predominate.

Note, that the northern mantle is mostly related to oils of Family III, which correspond to the Polier–Manacas (!) Petroleum System, described above. But, in the Rosario Sur mantle, the surface and well shows correspond to families I and II, which correlate with the source rocks of the Sierra del Rosario TSU of age $J_3^{km}-K_1^{ba}$ (Artemisa and Sumidero formations).

On this basis, it is stated that in the Rosario Sur mantle, there is a petroleum system associated with families I and II of Cuban crudes whose source rock would be the Artemisa and Sumidero formations (synchronous to the Veloz Group, Placetas TSU).

1.4.3.6 Petroleum Systems in Central Basin

The Central Basin is an area where the Ophiolites, the CVA and their associated "piggy back" basins are thrusted out of sequence on the sediments of the North American continental paleomargin. This assertion is made on the basis of the presence of small oil fields (Figs. 1.12 and 1.14) in the volcanic rocks, indicating the presence of source rocks of the North American continental paleomargin in depth.

The biomarker studies show the presence of families II and III of Cuban crudes in the Central Basin. In Sect. 4.2 (Fig. 1.13), it was argued that the source rocks that correlate with these families are those of the $J_3^{km}-K_1^{ba}$ and $K_1^h-K_2^t$ intervals, respectively. In the figure, it is observed that the source rocks of the Placetas TSU are the ones that best correlate with the oil fields of Central Basin, Cifuentes and Ronda, for Family II, and Carmita, for Family III. Based on the fact that west of the Central Basin the Placetas TSU (Figs. 1.4 and 1.14) outcrop, two active petroleum systems are proposed in the Central Basin: one related to Family II, whose source rock is Veloz Group ($J_3^{km}-K_1^{ba}$) and, another, related to Family III, whose source rock is the Carmita Formation (K_2^{c-t}).

The great majority of the elements of these systems are hidden in depth under the over thrusted mantles of the Ophiolites and the CVA and, to date, they have not been reached by drillings. However, taking into consideration that the crudes of the Central Basin show signs of dismigration or tertiary migration (see Sect. 2.1, concepts of migration), it is assumed that these have escaped from accumulations in reservoirs of the North American continental paleomargin in depth. These reservoirs and deep seals would be the same described in the petroleum systems Veloz Group–Veloz Group (!) (in the case of Family II) and Carmita–Ophiolites (!) (for Family III).

It is necessary to point out that the oils from Central Basin are the most mature in Cuba and, their commercial qualities classify them as light (those not altered by secondary processes) and medium (those affected by secondary processes), indicating that the source rocks that generated them reached high levels of thermal maturity. These data indicate that deep accumulations in traps related to the North American continental paleomargin (Placetas TSU) store mature crude of high commercial qualities, unlike the NCOS, where crude oils are of low maturity and, therefore, of low commercial quality.

1.4.3.7 Petroleum Systems in the Southern Petroleum Province

The petroleum systems of the Southern Province of Cuba have a hypothetical category (see Sect. 2.1, name of a petroleum system), as today the active source rock is not known, despite several rock eval studies carried out to samples from wells and outcrops. However, the hydrocarbon shows in the southern half of Cuba (Fig. 1.12) are direct index of the presence of active source rocks.

The studies of biomarkers of oil samples revealed the presence of families II and III of Cuban crude oil in the onshore part of the Southern Province (Fig. 1.12), suggesting the presence of petroleum systems similar to those defined for the Central

Basin. In the shows of the offshore part of this province (Ana María and Guacanayabo gulfs), the Oleanano biomarker was found, which comes from angiosperms. As the angiosperms arose in the Upper Cretaceous, this parameter is an index of the age of the source rock (Upper Cretaceous or younger). This indicates petroleum systems different from those known onshore.

1.5 Conclusions

1. The results of the application of the exploratory method of petroleum systems in Cuba have allowed obtaining exploratory criteria applicable to all geological scenarios.
2. In ,Cuba there are four stratigraphic intervals of source rocks (Middle Jurassic, Oxfordian Upper Jurassic, Tithonian Upper Jurassic–Barremian Lower Cretaceous, and Aptian Lower Cretaceous–Turonian Upper Cretaceous).
3. Three genetic oil families (I, II, and III) were identified. Families I and II were generated by the same level of source rock (Tithonian Upper Jurassic–Barremian Lower Cretaceous) product to facial variations. Family III was generated by the source rocks of the Aptian Lower Cretaceous–Turonian Upper Cretaceous.
4. The presence of oils from Family III is an unequivocal index of the presence in the subsoil of the tectonic mantle of the Lower Cretaceous upper part–Upper Cretaceous lower part (K^{ap-ce}).
5. The oils of Family I present better physical–chemical characteristics when they are found in reservoirs of the tectonic mantle (K^{ap-ce}), than when they are in the Upper Jurassic–Lower Cretaceous mantle, which also helps to define the presence of the first mantle in the subsoil.
6. The results of the exploratory method of Petroleum Systems indicate that in the Central Basin, there are mature oil fields of high commercial quality in reservoirs of the Placetas TSU buried under the tectonic mantle of the Ophiolites and the Cretaceous Volcanic Arc.
7. Biomarker data indicate that the petroleum systems identified in the Northern Province (onshore) can coexist in the basins of southern Cuba with other systems associated with source rocks of Late Cretaceous or Tertiary age (offshore), according to the presence of Oleanano in samples from the Guacanayabo Gulf.

References

Arellano J (2001) Las rocas generadoras de hidrocarburos del Jurásico relacionadas con la apertura del golfo de México. Segundo seminario bilateral Cuba – México sobre la geología del petróleo del sureste del Golfo de México. Villahermosa, Tabasco 12–14 de noviembre 2001

Cobiella Reguera JL (2000) Jurassic and Cretaceous geological history of Cuba. Int Geol Rev 42:594–616

Cobiella Reguera JL (2008) Reconstrucción palinspástica del paleomargen mesozoico de América del Norte en Cuba occidental y el sudeste del Golfo de México. Implicaciones para la evolución del SE del Golfo de México, Revista Mexicana de Ciencias Geológicas 25(3):382–401

Cobiella Reguera JL and Olóriz F (2009) Oxfordian–Berriasian stratigraphy of the North American paleomargin in western Cuba: Constraints for the geological history of the proto-Caribbean and the early Gulf of Mexico. In: Bartolini C, Román Ramos JC (eds) Petroleum systems in the southern Gulf of Mexico. AAPG Memoir 90, pp 421–451

Delgado López O (2003) Geoquímica de los Sistemas Petroleros presentes en Cuba Occidental. Tesis de M.Sc., Instituto Superior Politécnico José Antonio Echevarría, La Habana, Cuba, 147 p

González GR, Holguín QN (2001) Las rocas generadoras de Méxicol. Boletín de la Asociación Mexicana de Geólogos Petroleros XLIX(1–2):16–30

Guzmán Vega M, Castro Ortiz L, Román Ramos J, Morales L, Valdés L, Vázquez Covarrubias E, Ziga Rodríguez G (2001) El Origen del petróleo en las subprovincias mexicanas del Golfo de México, México. Boletín de la Asociación Mexicana de Geólogos Petroleros XLIX(1–2):31–46

Iturralde Vinent M (1996) Introduction to Cuban geology and geophysics. In: Iturralde-Vinent M (ed) Ofiolitas y arcos volcánicos de Cuba: International Geological Correlation Programe Project 364, Geological correlation of ophiolites and volcanic arcs in the Circumcaribbean realm, Miami, Florida, pp 3–35

López Corzo O, Perera C, Pérez Y, Otero R, García N, Menduiña R (2008) Informe final del pozo Angelina 100. Centro de Investigaciones.del Petróleo, La Habana (Informe Interno), p 14

Magoon LB, Beaumont EA (1999) Petroleum systems. In: Beaumont EA, Foster NH (eds) Exploring for oil and gas traps. American Association of Petroleum Geologist, Oklahoma, pp 3.1–3.34

Magoon LB, Dow WG (1994) Petroleum system-from source to trap. AAPG Memoir 60, pp 3–24

Marton G, Buffler R (1999) Jurassic–Early Cretaceous tectono-paleogeographic evolution of the southeastern Gulf of Mexico Basin. In: Mann P (ed) Caribbean basins: sedimentary basins of the World, vol 4, pp 63–91

Moretti I, Tenreyro R, Linares E, López Rivera JG, Letouzey J, Magnier C, Gaumet F, Lecomte J, López Quintero JO, Zimine S (2003) Petroleum systems of the Cuban northwest offshore zone. In: Bartolini C, Buffler R, Blickwelde J (eds) The circum Gulf of Mexico and the Caribbean: hydrocarbon habitats, basin formation, and plate tectonics. AAPG Memoir 79, pp 675–696

Pérez Y, López Corzo O, Perera C, Medina A, Brey D, Pérez L, Castro O, Menduiña R (2007) Informe final del pozo Cantel 2000 (R). Centro de Investigaciones del Petróleo, La Habana (Informe Interno), p 26

Pindell JL (1994) Evolution of the Gulf of Mexico and the Caribbean. In: Donovan S, Jackson T (eds) Caribbean geology. An introduction: Kingston, Jamaica. The University of the West Indies Publishers' Association, pp 13–40

Pindell JL, Barrett S (1990) Geological evolution of the Caribbean Region: a plate tectonics perspective. In: Dengo G, Case J (eds) The Caribbean region. The geology of North America, H. Geological Society of America, Boulder, CO, 405–432 pp

Pszczółkowski A (1999) The exposed passive Margin of North America in western Cuba. In: Mann P (ed) Sedimentary basins of the world, vol 4. Caribbean Basins, p 93

Pszczolkowski A, Myczynski R (2003) Stratigraphic constraints on the Late Jurassic–Cretaceous paleotectonic interpretations of the Placetas belt in Cuba. In: Bartolini C, Buffler RT, Blickwede J (eds) The Circum-Gulf of Mexico and the Caribbean: hydrocarbon habitats, basin formation, and plate tectonics. AAPG Memoir 79, pp 545–581

Rocha Mello M, Trindade LA (1996) Nuevas metas exploratorias en cuencas latinoamericanas de aguas profundas: Cómo proceder con el concepto de sistema petrolero. Oil & Gas Jornal Revista Latinoamericana 2(1):25–31

Sánchez Arango JR, López S, Sorá A, Domínguez R, Toucet S, Rodríguez R, Socorro R, Juara M (2007) Sismo estratigrafía y estratigrafía secuencial en el offshore del noroeste de Cuba. X Simposio de Geofísica, Veracruz, México, p 32

Sánchez Arango JR, Socorro R, López S, Sorá A, Domínguez R, Toucet S (2003) Estratigrafía integrativa aplicada en la Zona Económica Exclusiva (ZEE) de Cuba en el sureste del Golfo de

México. In: IV Congreso Cubano de Geología y Minería, La Habana, Cuba, 24–28 de marzo 2003: (CD Memorias, ISBN 959-7117-11-8)

Sánchez Arango JR, Tenreyro R, López Rivera JG, López Quintero JO, Blanco S y Valladares S (2001) An approach to the stratigraphy of the Cuban area in the southeastern Gulf of México. In: IV Congreso Cubano de Geología, La Habana, Cuba, 19–23 marzo 2001: (CD Memorias, ISBN 959-7117-10-X)

Santamaría D, Horsfield B (2001) Tendencias de evolución térmica de la materia orgánica en el área marina de Campeche, México. Boletín de la Asociación Mexicana de Geólogos Petroleros XLIX(1–2):116–136

Schlager W, Buffler R, Angstadt D, Phair R (1984) Geologic history of the southeastern Gulf of Mexico. In: Buffler RT, Schlager W, Bowdler JL, Cotillon PH, Halley RB, Kinoshita H, Magoon LB, McNulty CL, Patton JW, Silva IP, Avello Suárez O, Testarmata M, Tyson RV, Watkins DK (1984) Initial reports of the deep sea drilling project, vol LXXVII, Universidad de California, E.U., pp 715–738

Tenreyro R, Sánchez Arango JR, Otero R, Toucet S, López Rivera JG (2001) Análisis sismo estratigráfico y secuencial en la ZEECGoM. IV Congreso Cubano de Geología, La Habana, Cuba, 19–23 marzo 2001: (CD Memorias, ISBN 959-7117-10-X)

Tissot BP, Welte DH (1984) Petroleum formation and occurrence. Springer-Verlag, New York

Valladares S, Álvarez J, Segura R, García R, Fernández J, Toucet S, Villavicencio B, Núñez C (1996) Atlas de Reservorios Carbonatados de Cuba. Centro de Investigaciones del Petróleo, La Habana (Informe Interno), p 247

Chapter 2
Results of Non-seismic and Non-conventional Exploration Methods in the Regions of Ciego de Ávila and Sancti Spiritus, Central Cuba

Manuel Enrique Pardo Echarte and Osvaldo Rodríguez Morán

Abstract In various geological situations, seismic data provide little or no information about whether a trap is loaded with hydrocarbons or not. In other cases, when the acquisition is difficult and extremely expensive, or the quality of the information is poor due to geology or unfavorable surface conditions, it is the non-seismic exploration methods and, in particular, the unconventional methods of exploration, the only ones that can provide information about subtle stratigraphic traps. In addition, it is well documented that the generality of hydrocarbon accumulations have leaks or microseepage, which are predominantly vertical, as well as that they can be detected and mapped using various non-conventional and not seismic methods of exploration. The benefits in the use of non-seismic and non-conventional exploration methods, integrated with geological data and conventional methods, translate into a better evaluation of prospects and exploration risk; such is the purpose of the present investigation. The geological task posed to the geological–geophysical processing and interpretation consisted in the mapping of possible new gaso-petroleum targets that will base the exploration in the Pina-Ceballos (Northeast of the Central Basin) and Sancti Spíiritus regions. In addition, an evaluation by recognition works of the *Redox Complex* of several of these possible new targets was envisaged. The mapping of the areas of interest was proposed based on the presence of a complex of indicator anomalies, mainly gravimetric, aeromagnetic, and airborne gamma spectrometric. To this end, the gravimetric and aeromagnetic field at 1:50,000 scale, the airborne gamma spectrometry at 1:100,000 scale and the Digital Elevation Models 90×90 m and 30×30 m of the territory were processed. The results indicate that the Pina oilfield anomalous complex is recognized, at least, in four other new localities, although with less areal extension; one of them is the Paraíso sector. Other deposits

M. E. Pardo Echarte (✉) · O. Rodríguez Morán
Centro de Investigaciones del Petróleo (Ceinpet), Churruca, no. 481, e/Vía Blanca y Washington,
El Cerro CP 12000, La Habana, Cuba
e-mail: pardo@ceinpet.cupet.cu

O. Rodríguez Morán
e-mail: ormoran2016@gmail.com

M. E. Pardo Echarte et al., *Non-seismic and Non-conventional Exploration Methods for Oil and Gas in Cuba*, SpringerBriefs in Earth System Sciences,
https://doi.org/10.1007/978-3-030-15824-8_2

and prospects such as Brujo, Ceballos, and Pina Sur have anomalous complexes similar to Pina's, but incomplete in some of their attributes. The same happens for other established interest sectors. From the use of the **Redox Complex**, the presence of hydrocarbons in the depth was established in different sectors with indicator anomalous complexes, many of them coinciding with seismic structures. These sectors are, in Ciego de Ávila: Pina Sur, Pina Sur SO, Oeste de Ceballos 1, Oeste de Ceballos 2 and Pina Oeste Norte, and; in Sancti Spíritus, Gálata 2.

Keywords Hydrocarbons exploration · Non-seismic and non-conventional exploration methods · Gravimetry · Aeromagnetometry · Airborne gamma spectrometry · **Redox complex** · Digital elevation model

2.1   Introduction

The seismic exploration is insurmountable to provide structural and stratigraphic information, as well as for the cartography and the obtaining of images of traps and reservoirs. However, in various geological situations, seismic data provide little or no information about whether a trap is loaded with hydrocarbons or not. In other cases, when the acquisition is difficult and extremely expensive, or the quality of the information is poor due to geology or unfavorable surface conditions, it is the non-seismic exploration methods and, in particular, the unconventional methods of exploration, the only ones that can provide information about subtle stratigraphic traps. Also, it is well documented that the generality of hydrocarbon accumulations has leaks or microseepage, which are predominantly vertical, as well as that they can be detected and mapped using various non-conventional and non-seismic methods of exploration. The surface expression of hydrocarbon microseepage can take many forms, which determine the development of various detection methods, both direct (hydrocarbon gas geochemistry) and indirect (**Redox Complex**), as well as non-seismic geophysical methods (gravimetry, magnetometry, electrical and airborne gamma spectrometry), morphometry, and remote sensing. The benefits in the use of non-seismic and non-conventional exploration methods, integrated with geological data and conventional methods, translate into a better evaluation of prospects and exploration risk; such is the purpose of the present investigation. Its general and specific objectives are to establish the perspective sectors for hydrocarbons in the study region (General Objective) and to identify the gaso-petroleum character or not of the seismic structures and non-seismic anomalous complex in the sectors of interest established, in addition to map the development areas of serpentine rocks in the Pina-Ceballos region (two Specific Objectives).

2.1.1   Geographic Location and General Characteristics

The study area includes the regions of Ciego de Ávila and Sancti Spíritus. The research area in the Ciego de Ávila region covers the territory of the oilfields and prospects, Pina, Brujo, Paraíso, Ceballos and adjoining areas, within the limits of Lambert Cuba North coordinates: X: 720000–745000 and Y: 235000–250000. The research area in the Sancti Spiritus region covers only two locations (Gálata 1–2 and El Pinto), where recognition profiles by the **Redox Complex** were made.

According to Martínez Rojas et al. (2007), the first successful hydrocarbon research in the Central Basin dates back to the 1950s when American companies discovered the Jatibonico (1954), Cristales (1955), and Catalina (1956) oilfields. The Jatibonico oilfield was discovered from gravimetric data, as well as the location of the Cristales fault; however, the Catalina structure was revealed from the first seismic investigations carried out in the region (1955–1956). In the 1960s, the Central Basin was the largest oil-producing region in the country, which brought with it an accelerated development of exploration works, which have been extended by more than 50 years of research and development.

The geological model for the study area, according to the same authors, considers reservoirs of the type tuffs, sandstones, and conglomerates of tuffs, associated with the Cretaceous Volcanic Arc, and its seal in clayey sequences of the synorogenic cretaceous and postorogenic coverage (Maastrichtian–Lower Eocene).

For the establishment of sectors of gaso-petroleum interest in this region, it has been considered essential to characterize the Pina oilfield anomalous complex (Fig. 2.1), based on the presence of a set of indicator anomalies that consider the following non-seismic attributes:

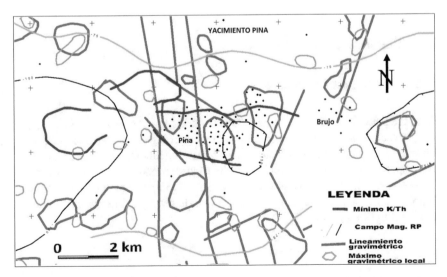

Fig. 2.1 Pina oilfield anomalous complex, based on the presence of a set of indicator anomalies

- Subtle local gravimetric maximum (from the first vertical derivative), which reflect positive structures (by the uplift of the more dense volcanic rocks), within the limits of the contours of the aeromagnetic field reduced to pole, in the interval between −240 and −150 nT, which defines, apparently, an area with similar geological–structural characteristics from the regional point of view (presumable uplift of volcanic rocks);
- Minimum of the K/Th ratio, with local maximums of U (Ra) in its periphery;
- Absence of geomorphic anomalies, and;
- Anomalous indications by the **Redox Complex**.

2.1.2 Geological Task

The geological task posed to geological–geophysical processing and interpretation consisted in the mapping of possible new gaso-petroleum targets that will base oil exploration in the Pina-Ceballos region (Northeast of Central Basin) and Sancti Spíritus. An evaluation through recognition work by the **Redox Complex** in several of these possible new targets was also foreseen. The mapping of the possible new gaso-petroleum targets was proposed from the presence of a complex of indicator anomalies, mainly gravimetric, aeromagnetic, and airborne gamma spectrometric, some of which would be evaluated by recognition profiles of the **Redox Complex**. To this end, the gravimetric and aeromagnetic field at 1:50,000 scale, the airborne gamma spectrometry at 1:100,000 scale and the Digital Elevation Models (DEMs) 90 × 90 m and 30 × 30 m of the territory were processed. The results of the recognition and evaluation works of the **Redox Complex** are also described.

2.2 Theoretical Framework

2.2.1 Geological Premises

From the petrophysical point of view (Density and Magnetic Susceptibility), the oil and gas sectors of interest are characterized by the lifting of the volcanic rocks, more dense, and magnetic in comparison with the remaining lithological varieties involved. This aspect considers both the regional point of view (trend of aeromagnetic reduced to pole values in the limits between −240 and −150 nT, which define a possible regional uplift of volcanic rocks) and local (subtle local gravimetric maximums on structural uplifts).

From the point of view of Surface Geochemistry, according to Price (1985), Schumacher (1996), Saunders et al. (1999), and Pardo Echarte and Rodríguez Morán (2016), the Geological Premises that base the application of the non-conventional geophysical–geochemical morphometric exploration methods are the following:

- The "Reducing Chimneys" are columns of mineralized rocks above the hydrocarbon deposits, which were modified by the vertical migration of these and/or by some other association of reduced species (metal ions) which are oxidized, by microbial action, to create a reducing environment.
- The main products of the microbial oxidation of hydrocarbons (CO_2) and the microbial reduction of sulfur (H_2S) drastically change the pH/Eh of the system.

Changes in pH/Eh result in changes in mineral stability:

- Precipitation of various carbonates.
- Decomposition of clays (consequently, increase the concentrations of silica and alumina).
- Precipitation of magnetite/maghemite, of iron sulfides (such as pyrrhotite and greigite) or coprecipitation of iron and/or manganese with calcite in carbonated cements on hydrocarbon deposits.

The morphometric, geophysical, and geochemical response to previous mineral stability changes is as follows:

- Secondary calcium carbonate mineralization and silicification result in denser and erosion-resistant surface materials (formation of positive residual geomorphic and maximum resistivity anomalies).
- The decomposition of the clay is responsible for the minimum radiation reported on the oil deposits: the potassium is leached from the system toward the edges of the vertical projection of the hydrocarbon deposit, where it precipitates resulting in a halo of high values. Thorio remains relatively fixed in its original distribution within insoluble heavy minerals; hence, minimum K/Th ratio are observed, surrounded by peaks on oil and gas deposits. In the periphery of the occurrences, maximums of U (Ra) are observed.
- The conversion of non-magnetic iron minerals (oxides and sulfides) into more stable magnetic varieties results in an increase in magnetic susceptibility, correlated with the minimum of Redox Potential, which justifies the integration of both techniques. Induced polarization anomalies are also observed.
- The arrival to the surface of the metallic ions contained in the hydrocarbons (V, Ni, Fe, Pb, and Zn, among others) condition the presence of a subtle anomaly of these elements in the soil and a slight change in coloration of the same, which is reflected, by anomalies of the spectral reflectance, facts that justify the integration of these techniques.

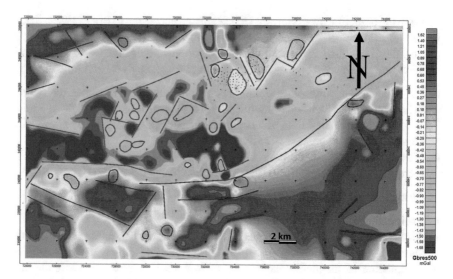

Fig. 2.2 Local gravimetric field (VD) of the Pina-Ceballos region. The local maximums and the tectonic alignments are indicated in gray-green outline. Drilling wells (black dots) are also observed

2.3 Materials and Methods

2.3.1 Information, Its Sources, and Processing

The materials used and their sources are the following: Mesh of the gravimetric and aeromagnetic field at 1:50,000 scale and, airborne gamma spectrometry (channels: It, U, Th, and K) at 1:100,000 scale of the Republic of Cuba (Mondelo et al. 2011). DEMs 90 × 90 m and 30 × 30 m from the Republic of Cuba (Sánchez Cruz et al. 2015 and https://lpdaac.usgs.gov/), respectively. Digital Map of the Oil Wells of the Republic of Cuba at 1:250,000 scale (Colectivo de Autores 2009).

The processing of the geophysical information was carried out using the software Oasis Montaj version 7.01.

The gravimetric field (Bouguer reduction, 2.3 t/m^3) was subjected to the regional-residual separation from the Upward Analytical Continuation (UAC) for heights of 500 and 2000 m, given by the order of depth of the possible oil and gas targets and, to the First Vertical Derivative (VD) (Fig. 2.2). The likely targets of interest are characterized by subtle local maximums within the regional minimum of the Central Basin. Different tectonic alignments were also mapped.

The aeromagnetic field underwent the reduction to pole (Fig. 2.3) and the VD, seeking to map the trend of values (between −240 and −150 nT) associated with the possible regional structural uplift of volcanic rocks and different tectonic alignments.

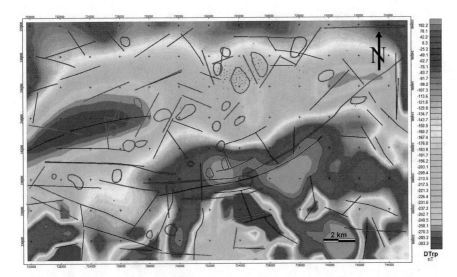

Fig. 2.3 Aeromagnetic field reduced to pole of the Pina–Ceballos region. The local gravimetric maximums and the gravimetric tectonic alignments appear indicated in gray-green outline; in black line, the magnetic alignments are indicated. Drilling wells (black dots) are also observed

The derivative of the magnetic field inclination was also determined (Fairhead et al. 2009), with the purpose of mapping the area of development (distribution) of the serpentinites (Fig. 2.4). In addition, an estimate was made of the depth of lying of these rocks in the limits of a regional fault to the east, at the height of the Ceballos town.

For the Airborne Gamma Spectrometry (AGS), the K/Th ratio was determined (Fig. 2.5), indicating the minimums surrounded by maximums (presumably linked to active zones of light hydrocarbon microseepage) and the local maximums of U (Ra) in their periphery.

For the elaboration of the map of morph alignments (Fig. 2.6), the maps of residual anomalies were used from the UAC at 500 m of the DEMs 90 × 90 m and 30 × 30 m.

2.4   Results

2.4.1   Results of Non-seismic Geophysical-Morphometric Methods

The results of the complex interpretation of non-seismic geophysical-morphometric methods in the Pina–Ceballos region are presented in Fig. 2.7.

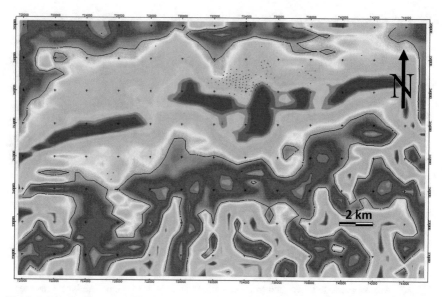

Fig. 2.4 Distribution areas of the serpentine rocks (orange-red-magenta color) of the Pina–Ceballos region, from the maximums of the derivative of the aeromagnetic field inclination. Drilling wells (black dots) are also observed

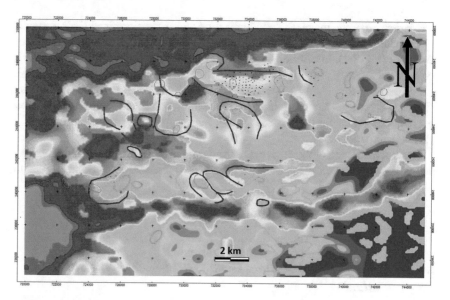

Fig. 2.5 K/Th ratio of the Pina–Ceballos region. The minimums of the K/Th ratio appear indicated in red line and, in pink line, the maximums of U (Ra). Drilling wells (black dots) are also observed

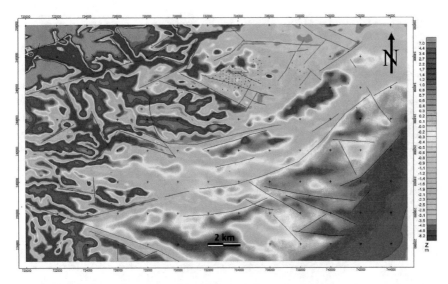

Fig. 2.6 Morphotectonic alignments of the Pina–Ceballos region (fine black lines) on the residual DEM 90 × 90 m at 500 m. Drilling wells (black dots) are also observed

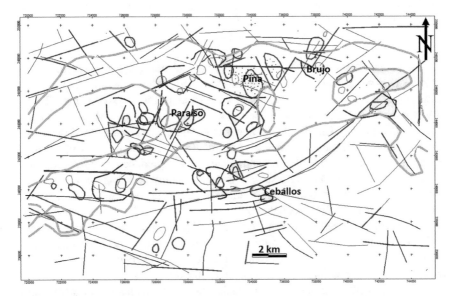

Fig. 2.7 Results of the complex interpretation of non-seismic geophysical-morphometric methods in the Pina–Ceballos region. In gray-green traces, the local gravimetric maximums and the gravimetric tectonic alignments are shown. In thick black lines, the aeromagnetic tectonic alignments are shown and, in light green, the regional trend of values between −240 and −150 nT of this field reduced to pole. In red line, the minimums of the K/Th ratio are indicated and, in pink, the increments of U (Ra). The morph alignments are represented in fine black traces. The measured recognition profiles of the **Redox Complex** are indicated in blue lines. Drilling wells (black dots) are also observed

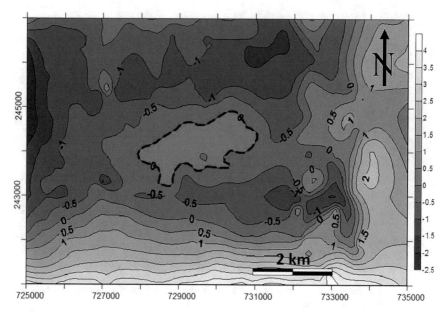

Fig. 2.8 Local gravimetric maximum (residual)—red trace, in the location of the Paraíso prospect, according to Pardo Echarte (2014)

The Pina oilfield anomalous complex (by non-seismic geophysical methods) has been recognized, at least, in four other new localities, although with less areal extension; one of them is the Paraíso sector. Images from the Paraíso sector (Figs. 2.8 and 2.9) argue very well for their perspectives, knowing, in addition, that it has a remarkable seismic structural uplift. These images represent a local low-amplitude gravimetric maximum and a satellite anomaly conveniently processed to represent the zones with presumable presence of hydrocarbons in the soil produced by microseepage.

Other deposits and prospects such as Brujo, Ceballos, and Pina Sur have anomalous complexes similar to the Pina oilfield but incomplete in some of its attributes. The same happens for other sectors of interest established in the present investigation.

According to the interpretation made (Fig. 2.7), possible targets of interest seem to be within the limits of the contours of the aeromagnetic field reduced to pole between −240 and −150 nT, defining a trend that characterizes, from the regional point of view, the possible structural uplift of volcanic rocks.

About the location of the serpentine rocks, the structural plane of the study region can be described, in general, as follows: The basin, properly speaking, is a central zone of southwest–northeast course of diminished fields, local gravimetric and reduced to pole aeromagnetic. Its flanks, western and eastern, are of increased (positive) values of the respective fields, with some dismemberment in some sectors product of the local tectonic activity (Figs. 2.2 and 2.3). On the flanks, the volcanics and its

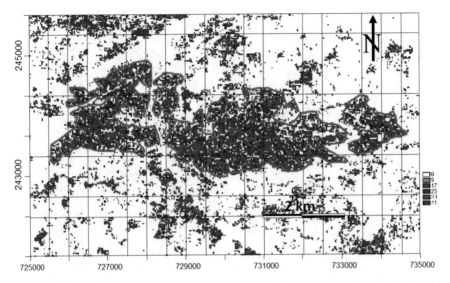

Fig. 2.9 Satellite image showing the area of possible hydrocarbon contamination in the ground (green trace) in the location of the Paraíso prospect, according to Pardo Echarte (2014)

oceanic basement are closer to the surface in relation to the basin (Fig. 2.4), where the sedimentary and sedimentary-volcanic thicknesses are much greater. The boundaries of the basin (on its flanks) are tectonic: the eastern one has a continuous form, and the western one is fractionally shaped according to the potential fields, although it is also revealed by a continuous regional morph alignment (Fig. 2.6).

In the Ceballos region, and to the west of the same prospect, an intense positive aeromagnetic anomaly is revealed within the basin (associated with the regional fault that limits it to the east), which seems to be caused by a local more superficial ophiolitic scale (Fig. 2.7). The depth of the serpentines (from the contour zero—zones of colors orange-red-magenta, Fig. 2.4) on the north edge of the fault is 1500–1600 m, contrasting with 700–1000 m, in its southern edge, which indicates that it dips toward the center of the basin. Hence, it is derived that the regional fault must be a major normal fault with approximately 500 m displacement (amplitude).

For the Sancti Spíritus region (Fig. 2.10), local gravimetric maximums are observed in sectors Gálata 1–2 (Fig. 2.11) and El Pinto (Fig. 2.12). In Fig. 2.13, the dashed line following the axis of the annular minimum around the local maximum of Gálata 1 suggests the limit of a possible gas cap (gas observed in the well). The anomaly of U (Ra) present is indicating that it is possibly a thermogenic and not a biogenic (methane) occurrence, as some researchers suggest.

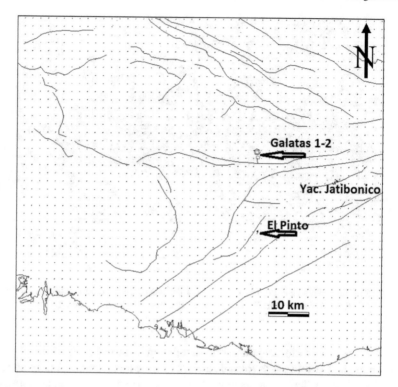

Fig. 2.10 Location of the Gálata 1–2 and El Pinto sectors in the Sancti Spíritus region

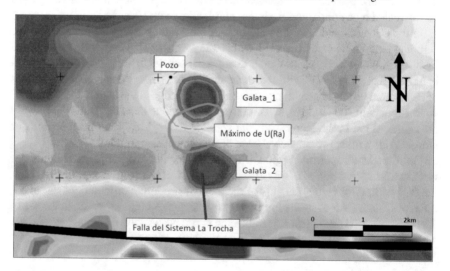

Fig. 2.11 Local gravimetric maximums (grayish-green trace) and U (Ra) anomaly (pink trace) in the Gálata 1–2 sector, Sancti Spíritus. In blue line, measured recognition profile of the *Redox Complex*

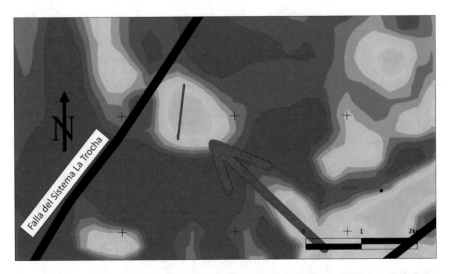

Fig. 2.12 Local gravimetric maximum (indicated by the red arrow) in the El Pinto sector, Sancti Spíritus. In blue line, measured recognition profile of the **Redox Complex**

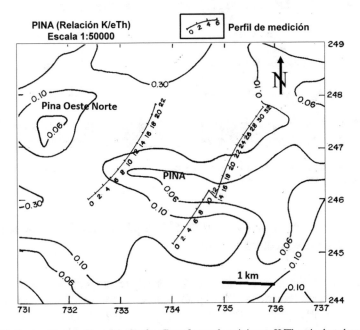

Fig. 2.13 Recognition profiles of the **Redox Complex** on the minimum K/Th ratio that characterizes the Pina oilfield, according to Pardo Echarte and Rodríguez Morán (2016)

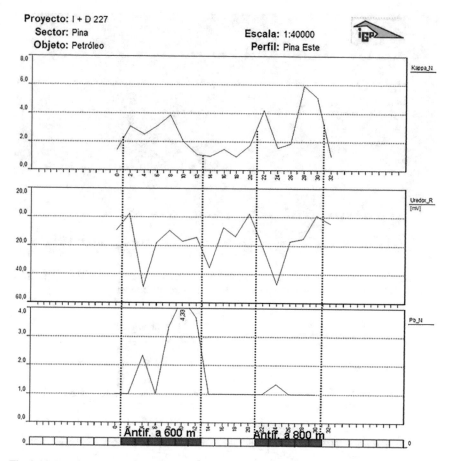

Fig. 2.14 Results of the works by the *Redox Complex* in the Pina Este profile, according to Pardo Echarte and Rodríguez Morán (2016)

2.4.2 Results of Non-conventional Geophysical-Geochemical Methods (**Redox Complex**)

The Pina oilfield has been recognized on land by two profiles of the **Redox Complex** (Fig. 2.13) (Pardo Echarte and Rodríguez Morán 2016), one of which (Pina Este) is represented in Fig. 2.14. This is an example of the response in the chemical elements (V, Ni, Fe, **Pb**, and Zn) that validates the presence of hydrocarbons in the depth (in this case, 600–800 m), on two nearby domes. The figure also shows the correlation of the normalized Pb increments with the increases of the normalized magnetic susceptibility (superficial magnetite dissemination) and the decrements of the reduced Redox Potential (superficial reducing environment).

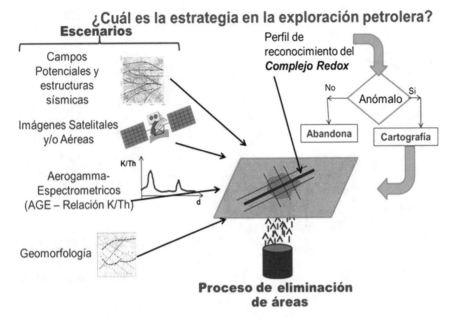

Fig. 2.15 Exploration Strategy for the application of the *Redox Complex*

The non-conventional geophysical–geochemical methods (*Redox Complex*) are applied following an exploration Strategy (Pardo Echarte and Rodríguez Morán 2016) which is represented in Fig. 2.15. According to this strategy, the areas of interest are recognized with a profile of the *Redox Complex* in cases of having, in whole or in part, the following information:

- Seismic: Structural uplifts of volcanic rocks, in the locality or very close to it.
- Gravimetry: Subtle local maximums (revealed, as a rule, in the VD), due to the uplift of the densest volcanic rocks; coincident or close to seismic uplifts.
- Aeromagnetometry: Regional trend of the field reduced to pole (between −240 and −150 nT) (possible higher part of the basin).
- Airborne gamma spectrometry: Minimums of the K/Th ratio and, in its periphery, local maximums of U (Ra).
- Morphometry: Subtle local maximums of the residual relief at 500 m, which, as a rule, are not observed.

The results of the *Redox Complex* works in the study region are described below, separated by campaigns (of two-four targets, each).

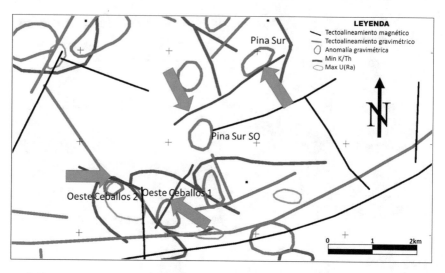

Fig. 2.16 Location of the measured *Redox Complex* profiles (blue lines) and signaling (with a blue arrow) of the anomalous intervals, during Campaign I

2.4.2.1 Results of Campaigns I and II (Ciego de Ávila Region)

During Campaign I, four targets were studied: Pina Sur, Pina Sur SO, Oeste de Ceballos 1 and Oeste de Ceballos 2, which are presented in Fig. 2.16 (a detail of Fig. 2.7).

The results of the works by the *Redox Complex* in each of the recognition profiles of the referred sectors are presented in Figs. 2.17, 2.18 and 2.19.

The main results achieved during this campaign are the following:

- The presence of hydrocarbons in the depth in the four sectors investigated is confirmed, based on the existence of increments greater than or equal to twice the background of the correlated contents of V and Ni (hydrocarbon indicator elements), coinciding with increases in Fe, Pb, and Zn.
- In all cases, a correlation between maximums of magnetic susceptibility with increases in the chemical elements (V and Ni) is observed. The behavior of the Redox Potential is not diagnostic, with maximums predominating; associated, presumably, with possible gaseous escapes.
- From the geochemical point of view (contents of V and Ni), the most contrast anomalous intervals are those of Pina Sur SO and Oeste de Ceballos 2, which exceed more than three times the background.
- The anomalous intervals are relatively narrow (500–900 m), which means small structures, more or less in the order of those mapped by the gravimetric maximums (0.4–0.5 km²).

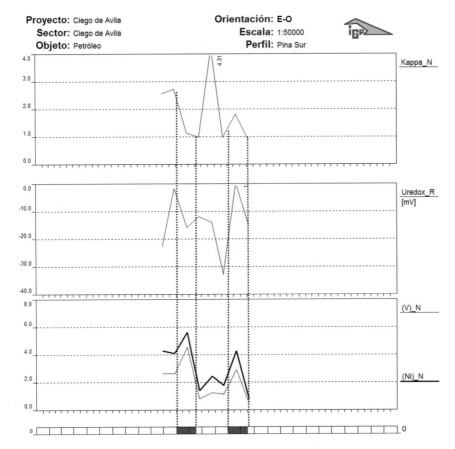

Fig. 2.17 Measurement profile Pina Sur SO-Pina Sur. The distance between measurement points is indicative

During Campaign II, three targets were studied: Pina Oeste Sur, Pina Oeste Norte, and América, which are presented in Fig. 2.20 (a detail of Fig. 2.7).

The results of the works by the **_Redox Complex_** in each of the recognition profiles in the referred sectors are presented in Figs. 2.21, 2.22 and 2.23.

The main results achieved during this campaign are the following:

- The possible* presence of hydrocarbons in the depth is confirmed in one of the three sectors investigated (Pina Oeste Norte-PON), based on the existence of increments equal to twice the background of the correlated contents of V and Ni (elements hydrocarbon indicators), coinciding with increases in Fe.
- In PON, a correlation is observed between the maximum of the magnetic susceptibility with an increase in the chemical elements (V and Ni). The behavior of the Redox Potential is not diagnostic, being a maximum; presumably associated with gaseous leaks.

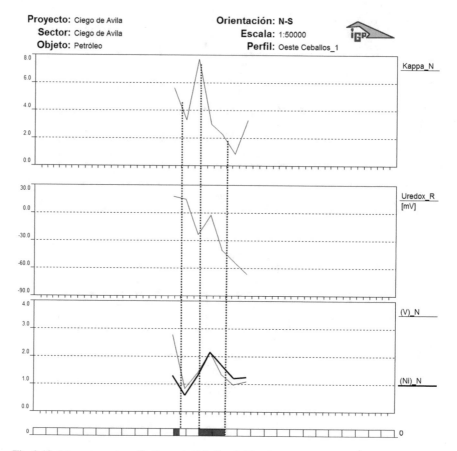

Fig. 2.18 Measurement profile Oeste de Ceballos 1. The distance between measurement points is indicative

- The anomalous interval in PON is relatively narrow (500–900 m) which implies a small structure, corresponding to the order of the mapped by the minimum of the K/Th ratio (Fig. 2.12).

* The term "possible" is justified by the relatively low anomalous contrast (two times the background) and the absence of correlation with increases in Pb and Zn.

2.4.2.2 Results of Campaign III (Sancti Spíritus)

During Campaign III, two targets were studied: Gálata 2 and El Pinto (Figs. 2.11 and 2.12). The results of the works by the **Redox Complex** in each of the recognition profiles in the referred sectors are presented in Figs. 2.23, 2.24 and 2.25.

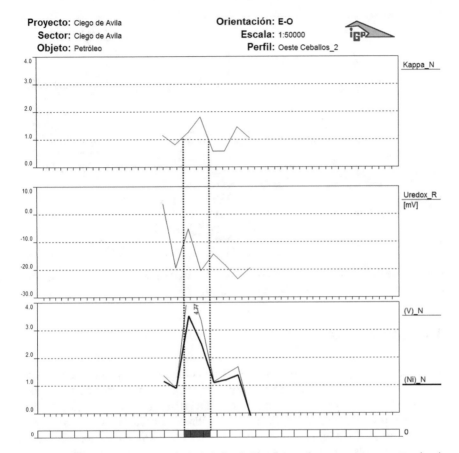

Proyecto: Ciego de Avila
Sector: Ciego de Avila
Objeto: Petróleo

Orientación: E-O
Escala: 1:50000
Perfil: Oeste Ceballos_2

Fig. 2.19 Measurement profile Oeste de Ceballos 2. The distance between measurement points is indicative

The main results achieved during this campaign were the following:

The possible* presence of hydrocarbons in the depth is confirmed in one of the two sectors investigated (Gálata 2), based on the existence of increments equal to twice the background of the correlated contents of V and Ni (hydrocarbon indicator elements), coincident with increments of Fe. There is a very narrow anterior anomalous zone before entering the main anomaly (local gravimetric maximum), possibly gaseous, analogous, perhaps, to that of Gálata 1.

* The term "possible" is justified by the relatively low anomalous contrast (two times the background) and the absence of correlation with increases in Pb and Zn.

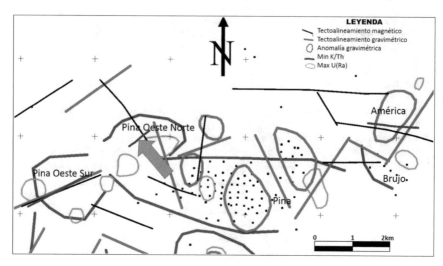

Fig. 2.20 Location of the measured *Redox Complex* profiles (blue lines) and signaling (with a blue arrow) of the anomalous intervals, during Campaign II

2.5   Conclusions

- The anomalous complex (by non-seismic geophysical-morphometric methods) of the Pina oilfield has been recognized, at least, in four other new localities, although with less areal extension; one of them is the Paraíso sector. Other deposits and prospects such as Brujo, Ceballos, and Pina Sur have anomalous complexes similar to the Pina oilfield, but incomplete in some of their attributes. The same happens for other sectors of interest established in the present investigation.
- According to the interpretation made, the possible targets of interest seem to be within the limits of the contours of the aeromagnetic field reduced to pole between −240 and −150 nT, which define a trend that characterizes, from the regional point of view, the possible structural uplift of the volcanic rocks.
- For the Sancti Spíritus region, local gravimetric maximums are observed in the Gálata 1–2 and El Pinto sectors. The annular minimum around the local maximum of Gálata 1 suggests the limit of a possible gas cap (gas observed in the well). The anomaly of U (Ra) present is indicating that it is possibly a thermogenic and not a biogenic (methane) occurrence, as some researchers suggest.

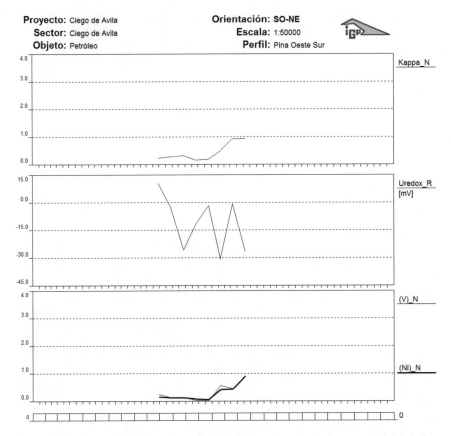

Fig. 2.21 Measurement profile Pina Oeste Sur. The distance between measurement points is indicative

- Based on the use of non-conventional geophysical–geochemical exploration techniques of indirect detection (**Redox Complex**), the presence of hydrocarbons in the depth is established in different sectors with anomalous indicator complexes, many of them coinciding with seismic structures. These sectors are in Ciego de Ávila: Pina Sur, Pina Sur SO, Oeste de Ceballos 1, Oeste de Ceballos 2, and Pina Oeste Norte and in Sancti Spíritus, Gálata 2.

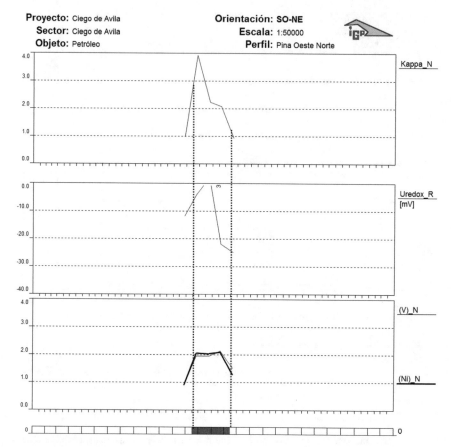

Fig. 2.22 Measurement profile Pina Oeste Norte. The distance between measurement points is indicative

- From the geochemical point of view (contents of V and Ni), the most contrast anomalous intervals correspond to the sectors of Pina Sur SO and Oeste de Ceballos 2, which exceed more than three times the background. In general, all the anomalous intervals are relatively narrow (500–900 m) which implies small structures, more or less in the order of those mapped by the gravimetric maximums (0.4–0.5 km^2).

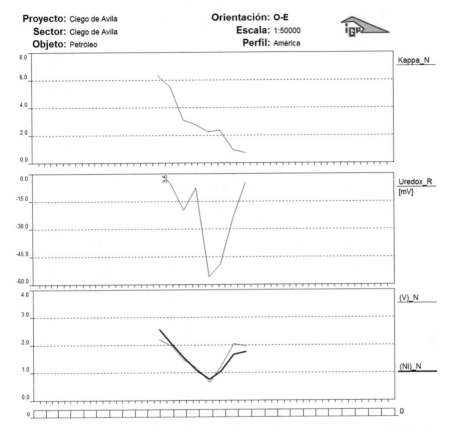

Fig. 2.23 Measurement profile América. The distance between measurement points is indicative

- In the Ceballos region and to the west of the same prospect, an intense positive aeromagnetic anomaly within the basin (associated with a regional fault that limits it to the east) seems to be caused by a more superficial local ophiolitic scale. The depth of the serpentines on the northern edge of the fault is 1500–1600 m and, 700–1000 m, on its southern edge, indicating that they dip toward the center of the basin. Hence, it is derived that the regional fault must be a major normal fault with approximately 500-m displacement (amplitude).

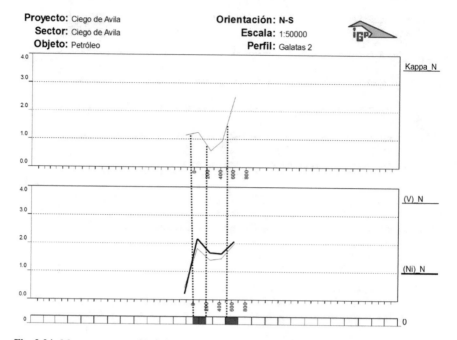

Fig. 2.24 Measurement profile Gálata 2. The distance between measurement points is indicative

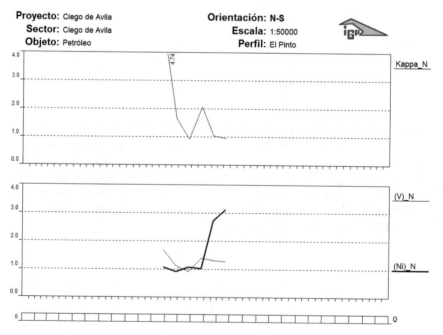

Fig. 2.25 Measurement profile El Pinto. The distance between measurement points is indicative

References

Colectivo de Autores (2009) Mapa Digital de los Pozos Petroleros de la República de Cuba a escala 1:250000. Inédito. Centro de Investigaciones del Petróleo, La Habana

Fairhead JD, Salem A, Williams SE (2009) Tilt-depth: a simple depth-estimation method using first order magnetic derivatives. Search and Discovery Article #40390 (Adapted from poster presentation at AAPG International Conference and Exhibition, Cape Town, South Africa, October 26–29, 2008)

Martínez Rojas E, Iparraguirre JL y otros (2007) Configuración tectono-estructural de la región noreste de Cuenca Central: una visión preliminar. Memorias del I Congreso de Petróleo y Gas (PETROGAS' 2007), 6 pp

MED 30 × 30 m de la República de Cuba (https://lpdaac.usgs.gov) USGS, NASA

Mondelo F, Sánchez Cruz R y otros (2011) Mapas geofísicos regionales de gravimetría, magnetometría, intensidad y espectrometría gamma de la República de Cuba, escalas 1:2,000,000 hasta 1:50,000. Inédito. IGP, La Habana, 278 p

Pardo Echarte ME (2014) El Complejo Redox en la Exploración de Petróleo y Gas. Curso de Postgrado. Inédito. Archivo Ceinpet, La Habana, 35 pp

Pardo Echarte ME, Rodríguez Morán O (2016) Unconventional methods for oil & gas exploration in Cuba. Springer Briefs in Earth System Sciences. https://doi.org/10.1007/978-3-319-28017-2

Price LC (1985) A critical overview of and proposed working model for hydrocarbon microseepage. US Department of the Interior Geological Survey. Open-File Report 85-271

Sánchez Cruz R, Mondelo F y otros (2015) Mapas Morfométricos de la República de Cuba para las Escalas 1:1000000–1:50000 como apoyo a la Interpretación Geofísica. Memorias VI Convención Cubana de Ciencias de la Tierra, VIII Congreso Cubano de Geofísica. http://www.cgiar-csi.org/data/srtm-90m-digital-elevation

Saunders DF, Burson KR, Thompson CK (1999) Model for hydrocarbon microseepage and related near-surface alterations. AAPG Bull 83(1):170–185

Schumacher D (1996) Hydrocarbon-induced alteration of soils and sediments. In: Schumacher D, Abrams MA (eds) Hydrocarbon migration and its near-surface expression. AAPG Memoir 66, pp 71–89

Chapter 3
Results of Non-seismic Exploration Methods in the Habana–Matanzas Region, Cuba

Manuel Enrique Pardo Echarte

Abstract A version of the geo-structural mapping of the study region, based on the gravitational-magnetic data and the mapping of sectors of gasopetroliferous interest, linked to the conventional oil of the Placetas Tectonic-Stratigraphic Unit, is offered, based on the presence of a complex of indicator anomalies. The source materials are: the gravimetric (Bouguer reduction 2.3 t/m^3) and aeromagnetic (reduced to pole) maps at a scale of 1:50,000; airborne gamma spectrometric maps (channels K, Th and U (Ra)) at scale 1:100,000; the Digital Elevation Model (90 × 90 m) and; a result map of remote sensing for the search of perspective gasopetroliferous sectors in the region of Guanabo-Seboruco. The processing consisted in the regional-residual separation of the gravimetric and morphometric fields, the calculation of the derivative of the magnetic field inclination and of the ratio of the K/Th spectrometry channels. The indicator anomalous complex considers the following attributes: low-amplitude local gravimetric maximums; minimum K/Th ratio and local maximums of U (Ra) at its periphery; local maximums of residual relief and remote sensing anomalies. The result is a complementary non-seismic information, essential for the necessary geological–geophysical–geochemical–morphometric integration of the territory. As a result of the geo-structural mapping from the gravi-magnetic data, a wide distribution of the Zaza Terrain (volcanic + ophiolites) can be observed in the study region. The main structural depressions are concentrated along a latitudinal strip that covers the following locations (from east to west): Southwest of Matanzas Bay, Ceiba Mocha, Aguacate, Bainoa, Tapaste, Cuatro Caminos, Managua, and Santiago de las Vegas. Based on estimates of the reduced at the pole magnetic field, the depth at the top of a target located to the west of Bainoa, within this strip, is 1350–1450 m, which gives an idea of its sedimentary thickness. The results of the integrated prospective cartography consider, in a first level of perspective, three localities (Boca de Jaruco, Jibacoa del Norte and Este de Aguacate) where all the studied anomalies (attributes),

M. E. Pardo Echarte (✉)
Centro de Investigaciones del Petróleo (Ceinpet), Churruca, no. 481, e/Vía Blanca y Washington, El Cerro CP 12000, La Habana, Cuba
e-mail: pardo@ceinpet.cupet.cu

© The Author(s), under exclusive license to Springer Nature Switzerland AG 2019　　69
M. E. Pardo Echarte et al., *Non-seismic and Non-conventional Exploration Methods for Oil and Gas in Cuba*, SpringerBriefs in Earth System Sciences, https://doi.org/10.1007/978-3-030-15824-8_3

with the exception of the morphometric ones, appear. In a second level of perspective, the localities that correspond to the combination of two types of different anomalies (11 localities) are considered.

Keywords Non-seismic exploration · Gravimetry · Airborne magnetics · Airborne gamma spectrometry · Morphometry · Remote sensing · Geological-structural mapping · Integrated prospective mapping · Unconventional methods of hydrocarbon exploration

3.1   Introduction

3.1.1   Geographical Location and Geological Characteristics

The study area belongs to the Habana–Matanzas region, between Morro-Cabañas and Matanzas Bay, within the approximate limits of Lambert Cuba North coordinates X: 361,000–451,000; Y: 339,000–375,000.

The Habana-Matanzas region, between Morro-Cabañas and Varadero-Cárdenas, is also known as the North Cuban Oil Strip (NCOS). According to the Colectivo de Autores (2009), the NCOS covers the coastal strip of the provinces of Habana and Matanzas, including the adjacent aquifer, about 5 km wide and 150 km long, where most and largest oilfields of the country have been discovered, although it is possible that it extends even more toward the west and the east.

In this strip, the two largest oilfields in Cuba (Varadero and Boca de Jaruco) were discovered at the end of the 1960s, although their current extension was established at the 1990s, when the participation of foreign companies carried out various 2D and 3D seismic campaigns. These revealed a train of structures along the coast whose exploratory drilling confirmed the oilfields of Puerto Escondido–Canasí, Yumurí–Seboruco and the west extension of Varadero deposit (Varadero Oeste).

Subsequently, other oilfields have been discovered, to name a few: Santa Cruz, Tarará, Bacuranao, Jibacoa, Habana del Este, and Morro-Cabañas. The density of the oils found fluctuates between 11° and 14° API, being achieved from the drilling directed with large angles (horizontal) wells with stable inputs of the order of up to 4000 barrels/day.

The subsurface geology of these sectors is one of the best known in the country. According to the aforementioned source, it is characterized by several levels of ramp folds against reverse faults of rocks of the North American Continental Margin and its coverage. These folds have probably been further complicated by shear accidents. Deformed rocks cover an age range that goes from the Jurassic to the Eocene. Intensely fractured and leached limestones covered by a clayey seal from the Paleocene to the Eocene represent the reservoirs; they have their analogs in the rock outcrops of the North American Continental Margin in Central Cuba (Tectono-Stratigraphic Unit [TSU] Placetas). The stacking of several anticline ramp folds is one of the main exploratory targets. These conform antiforms that are difficulty mapable by the seismic. The poor image is the main obstacle for the development of exploratory works; only the one directly related to the envelope of the folds is observed as a horizon with high dynamic definition. Hence, the scientific problem posed to research is the need to integrate all the geological–geophysical–geochemical–morphometric information of the territory to increase the geological effectiveness of the exploration and reduce its risks.

3.1.2 Geological Task

The geological task posed to the geophysical–morphometric processing and interpretation of the study region, and General Objective of the research, consisted of establishing the sectors of gaso-petroleum interest linked to the conventional oil and gas of the TSU Placetas, based on the presence of a complex of indicator geophysical–morphometric anomalies. As a specific objective, it was proposed to carry out the geological–structural cartography of the territory from the potential fields. For such purposes, the gravimetric and aeromagnetic field at 1:50,000 scale, the airborne gamma spectrometry (AGS) at scale 1:100,000 and the Digital Elevation Model (DEM) 90 × 90 m were processed. For the final integration, the results of the Remote Sensing (RS) in the area of Guanabo–Seboruco (Jiménez de la Fuente 2017) were used as complementary information.

3.2 Theoretical Framework

3.2.1 Geological Premises

From the petrophysical point of view, the table below (Table 3.1) shows the results of the Petrophysical Generalization of Matanzas Province, which must be very similar to that of Habana. According to the same, by the high density of the K-J carbonates, the elevations of the top of them can cause local gravimetric maximums of low amplitude. Concerning to magnetic susceptibility, there are no significant contrasts in the aforementioned section (with the exception of effusive and ultrabasic), so only anomalies related to these targets are expected.

From the point of view of surface geochemistry, according to Price (1985), Schumacher (1996), Saunders et al. (1999) and Pardo Echarte and Rodríguez Morán (2016), the geological premises that base the application of the non-conventional geophysical–geochemical–morphometric exploration techniques are the following:

- The "Reducing Chimneys" are columns of mineralized rocks above the hydrocarbon deposits, which were modified by the vertical migration of these and/or by some other association of reduced species (metal ions) which "oxidize," by microbial action, to create a reducing environment.

Table 3.1 Petrophysical generalization of Matanzas Province (Pardo Echarte 2016)

Geol. Frm. or Geol. unit	Age	Lithology	Density (t/m³)				Magnet. Suscept. (10⁻³ SI)			
			Amount samples	Min	Max	Average	Amount samples	Min	Max	Average
Jaimanitas	Pleistoc.	Coral limest.	48	1.70	2.44	2.12	–	–	–	–
Güines	N_1^{1-2}	Limest.	1094	1.96	2.98	2.45	178	0	0.04	0.02
		Dolomites	632	2.71	2.85	2.80	–	–	–	–
Peñón	P_2^2	Calcium breccias	13	2.50	3.17	2.81	13	0.1	2.39	0.34
		Biocalcar. and marl	13	2.03	2.70	2.45	53	0	0.54	–
Perla	P_2^1	Marl	68	1.48	2.84	2.11	97	0	0.54	0.05
Vía Blanca	K_2^{cp-m}	Flysch terrigen.	499	1.96	2.92	2.38	15	0	0.56	–
Carmita	K_2^{cm-st}	Limest. and silicites	–	–	–	2.65	–	–	–	0.05
Amaro	K_2	Calcium conglom. breccias	32	2.44	2.70	2.58	32	0	0.34	0.16
Santa Teresa	$K_1^a-K_2^l$	Silicites and clays	17	1.85	2.63	2.32	5	0	5.29	–
Chirino	K_{1-2}	Efusive	–	–	–	2.57	–	–	–	8.71
		Tuff	–	2.45	2.60	–	–	0.16	18.30	5.43
Grupo Veloz	$J_3^l-K_1^v$	Limest.	152/62	1.92	2.70	2.28/2.57	214	0	0.16	0.032
		Sandst.	13	2.30	2.40	2.36	–	–	–	–
Constancia	$J_{1-2}-J_3$	Limest. and sandst.	–	–	–	2.65	–	–	–	–
Complejo Ofiolítico	Tr-J	Serpent.	7/2	2.20	2.70	2.37/2.63	11/-	0.05	24.67	8.69/-
		Ultrabasic	–	–	–	–	–	23.87	39.79	–

- The main products of the microbial oxidation of hydrocarbons (CO_2) and the microbial reduction of sulfur (H_2S), drastically change the pH/Eh of the system.

Changes in pH/Eh result in changes in mineral stability:

- Precipitation of various carbonates.
- Decomposition of the clays (consequently, increase the concentrations of silica and alumina).
- Precipitation of magnetite/maghemite, of iron sulfides (such as pyrrhotite and greigite) or coprecipitation of iron and/or manganese with calcite in carbonated cements on hydrocarbon deposits.

The morphometric, geophysical, and geochemical response to previous mineral stability changes is as follows:

- Secondary mineralization of calcium carbonate and silicification results in more dense and erosion-resistant surface materials (formation of geomorphic anomalies and resistivity maximums).
- The decomposition of the clay is responsible for the minimum radiation reported on the oil deposits: potassium is leached from the system toward the edges of the vertical projection of the hydrocarbon deposit, where it precipitates resulting in a "halo" of high values. Thorio remains relatively fixed in its original distribution within insoluble heavy minerals; hence, minimum K/Th ratios are observed, surrounded by peaks on oil and gas deposits. In the periphery, maximums (increments) of U (Ra) are observed.
- The conversion of non-magnetic iron minerals (oxides and sulfides) into more stable magnetic varieties results in an increase in magnetic susceptibility, correlated with the minimum of the Redox Potential, which justifies the integration of both techniques. Induced polarization anomalies are also observed.
- The arrival to the surface of the metallic ions contained in the hydrocarbons (V, Ni, Fe, Pb, and Zn, among others) condition the presence of a subtle anomaly of these elements in the soil. These anomalies are the main indicator of the presence of hydrocarbons in the depth.

3.3   Materials and Methods

3.3.1   Information and Its Sources

The materials used and their sources are the following:

- Meshes of the gravimetric and aeromagnetic field at scale 1:50,000 and, airborne gamma spectrometry (channels: It, U, Th and K) at scale 1:100,000 of the Republic of Cuba (Mondelo et al. 2011).
- The DEM (90 × 90 m) used in this work was taken from Sánchez Cruz et al. (2015), with a source at: http://www.cgiar-csi.org/data/srtm-90m-digital-elevation.

- Results of Remote Sensing (RS) for the search of prospective gaso-petroleum sectors in the region of Guanabo-Seboruco (Jiménez de la Fuente 2017).
- Digital Geological Map of the Republic of Cuba at scale 1:100,000. Colectivo de Autores (2010).

The processing of the geophysical–morphometric information was carried out using the software Oasis Montaj version 7.01.

3.4   Results and Discussion

3.4.1   Processing and Interpretation of Information

The gravimetric field (Bouguer reduction, 2.3 t/m^3) was subjected to regional-residual separation from the Upward Analytical Continuation (UAC) for the heights of 500, 2000 and 6000 m, given by the order of depth of the possible gaso-petroleum targets and the seismic study. However, the establishment of positive local anomalies of low amplitude, with an order of depth of 500–1000 m, was made from the first vertical derivative (VD) (Fig. 3.1). In the local plane of this figure, the maximums are associated with the presence of the Zaza Terrain (volcanic + ophiolite) and, the minimum, with structural depressions. The results of gravimetric mapping (alignments and local maximums) are presented in Fig. 3.2.

In aeromagnetometry, the ability to map geological–structural features is reinforced by the ability to map anomalies of small amplitude. Intrusive bodies (granitoids) and protrusions (ophiolites) can often be distinguished directly based

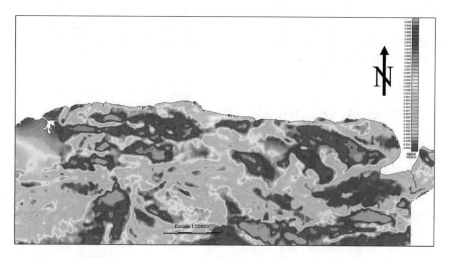

Fig. 3.1 Local gravimetric field of the study region from the VD

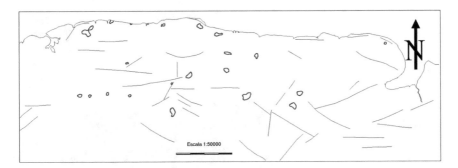

Fig. 3.2 Results of gravimetric mapping (alignments and local maximums)

on magnetic observations. According to Fairhead et al. (2009), the derivative of the total magnetic field inclination and its total horizontal derivative is useful for structural cartography and exploration. The derivative of the field inclination is defined as follows:

$$\text{TDR} = \arctan\left(\frac{\text{VDR}}{\text{THDR}}\right) \tag{3.1}$$

where VDR and THDR are the first vertical derivative and the total horizontal derivative, respectively, of the total magnetic intensity T, reduced to the pole.

$$\text{VDR} = \frac{\mathrm{d}T}{\mathrm{d}z\text{R}} \tag{3.2}$$

$$\text{THDR} = \sqrt{\left(\frac{\mathrm{d}T}{\mathrm{d}x}\right)^2 + \left(\frac{\mathrm{d}T}{\mathrm{d}y}\right)^2} \tag{3.3}$$

The total horizontal derivative of the slope derivative is defined as follows:

$$\text{HD\_TDR} = \sqrt{\left(\frac{\mathrm{d}\text{TDR}}{\mathrm{d}x}\right)^2 + \left(\frac{\mathrm{d}\text{TDT}}{\mathrm{d}y}\right)^2} \tag{3.4}$$

The axis of the chains of maximums of this attribute coincides with structural limits or tectonic alignments.

The derivation of the inclination of the magnetic field reduced to the pole allows the estimation of the depth up to the top of magnetic targets, in our case, presumable ophiolitic bodies.

The aeromagnetic field was subjected to the pole reduction (Fig. 3.3). In the plane of the figure, the maximums are associated to the presence of the Zaza Terrain (volcanic + ophiolite) and, the minimums, to structural depressions. The results of magnetic mapping (alignments) are presented in Fig. 3.4.

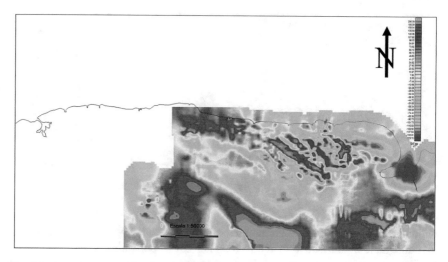

Fig. 3.3 Aeromagnetic field reduced to pole of the study region

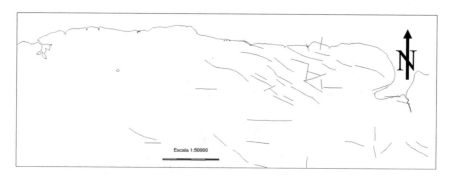

Fig. 3.4 Results of magnetic cartography (alignments)

For the airborne gamma spectrometry (AGS), the K/Th ratio was determined, with the purpose of indicating the minimums, presumably linked to active zones of light hydrocarbons microseepage (Fig. 3.5). The results of the AGS mapping (minimums of the K/Th ratio and local maximums of U (Ra) in its periphery) are presented in Fig. 3.6.

The DEM (90 × 90 m) underwent the regional-residual separation from the UAC at 500 m, according to the author's experience, to indicate the local maximums, linked to the light carbonization and subsurface silicification processes that take place on the active light hydrocarbons microseepage (Fig. 3.7). The results of the morphometric cartography (local geomorphic alignments and maximums) are presented in Fig. 3.8.

In RS, according to Jiménez de la Fuente (2017), the search and processing of the images corresponding to the Guanabo–Seboruco region, included between the coordinates X: 382000-446300 and Y: 338000-372000, from the scenes AST_L1B_00303132005161351_20081014083203_32336 and

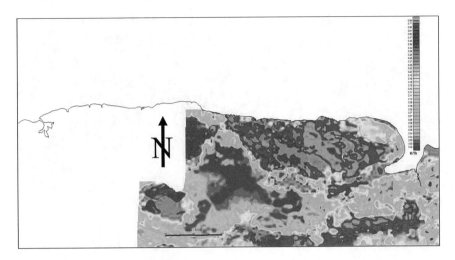

Fig. 3.5 K/Th ratio of the study region

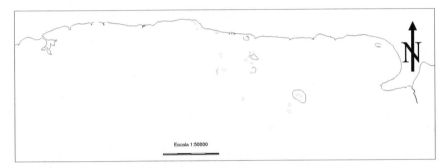

Fig. 3.6 Results of AGS cartography (minimums of the K/Th ratio, in red and local maximums of U (Ra), in pink)

AST_L1B_00303012003160859_20081014120752_2085 of the Aster sensor. The processing consisted in the calculation of ratios of bands that responded to lithological and mineralogical changes produced by possible accumulations of hydrocarbons in depth. The ratios of bands 2/1 and 4/9 allowed to map areas with possible alterations by ferric oxides and carbonates, respectively. The area was divided into two zones, north and south, where anomalies were mapped following the criteria:

- Northern Zone (related to the main gaso-petroleum oilfields, minor anomalies): Maximum values of carbonate ratio and high values of the iron ratio not related to anthropic elements.
- Southern Zone (anomalies of larger dimensions): Low-medium values of carbonate ratio and high values of the iron ratio not related to anthropic elements.

The results of the RS mapping (RS anomalies) are presented in Fig. 3.9.

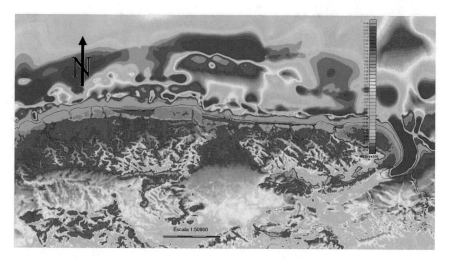

Fig. 3.7   Residual DEM (90 × 90 m) at 500 m of the study region

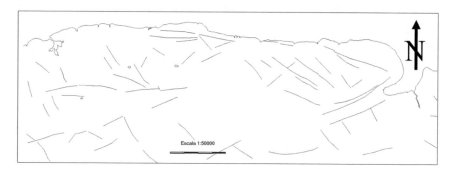

Fig. 3.8   Results of the morphometric cartography (geomorphic alignments and local maximums)

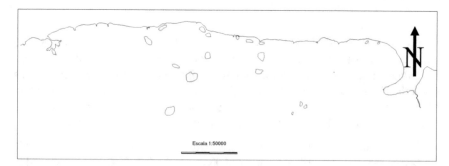

Fig. 3.9   Results of the RS cartography (anomalies)

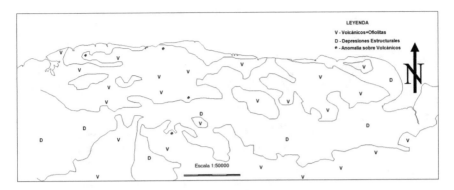

Fig. 3.10 Geological–structural cartography based on gravimetric data

The geophysical and morphometric alignments (fundamentally, regional tectonic dislocations) were plotted, fundamentally, from the VD map of the aeromagnetic and gravimetric fields and the residual DEM at 500 m, considering, basically, the chains of minimums and some of the maximums, linearity, bending, and interruption of the contours, as well as areas of high gradient thereof.

Likewise, the main structural depressions were recognized, considering a change in the characteristics of the field of magnetic VD (smoothing or flattening), coinciding with gravitational minimums.

The geophysical processing was carried out with two main purposes: geological –structural cartography (from the potential fields and morphometry (Figs. 3.10 and 3.11) and, integrated prospective cartography (from all geophysical fields and morphometry; Fig. 3.12). The first of them aims to reveal the main elements of the geological–structural picture from the gravimetric data (Fig. 3.10—cartography of the volcanic + ophiolites terrain and structural depressions). The second one (Fig. 3.12) has the purpose of establishing localities with perspectives for oil and gas deposits, considering the presence of a complex of indicator anomalies: local low-gravimetric maximums, coinciding with minimums of the K/Th ratio, local maximums of U (Ra) in the periphery, and local morphometric maximums. As complementary information, very sensitive, the presence of RS anomalies is considered.

For the purposes of the geological interpretation was used the geological map of Cuba at 1:100,000 scale of the IGP (Colectivo de Autores 2010).

As a result of the geological–structural cartography from the gravimetric data (Fig. 3.10), a wide distribution of the Zaza Terrain (volcanic + ophiolites) can be observed in the study region. The main structural depressions are concentrated along a strip of latitudinal direction that encompasses the following locations (from east to west): Southwest of Matanzas Bay, Ceiba Mocha, Aguacate, Bainoa, Tapaste, Cuatro Caminos, Managua, and Santiago de las Vegas. Based on the estimates from the magnetic field reduced to pole, the depth at the top of a target located west of Bainoa, within this strip, is 1350–1450 m, which gives an idea of the sedimentary thickness in it. Another important depression exists, of transversal course (N-S) to

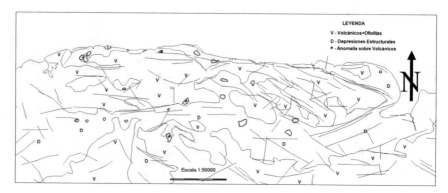

Fig. 3.11 Geological–structural cartography (with alignments) based on potential fields and morphometry

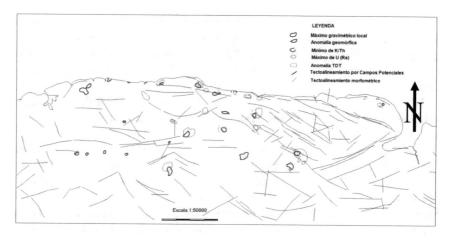

Fig. 3.12 Results of integrated prospective cartography

the previous strip, through the locality of San José de las Lajas. Another structural depression has a clear expression in the gravimetric minimum of Punta Rubalcava to the NNE of Matanzas.

The local gravimetric maximums that are located on the Zaza Terrain, having, therefore, a minor gaso-petroleum perspective, are located to the SE of Alamar, SE of Brisas del Mar, SW of San Antonio de Río Blanco, and SE of El Perú.

The results of the integrated prospective cartography (Fig. 3.12) consider, at a first level of prospectivity, three localities where all the studied anomalies appear, with the exception of the morphometric ones (local gravimetric maximum + minimum of K/Th + ratio) with maximums of U (Ra) in the periphery + RS anomalies). In a second level of prospectivity, the localities (eleven) correspond to the combination of two different types of anomalies. The anomalous localities for the two levels indicated are presented in Table 3.2.

Table 3.2 Anomalous localities separated by levels of prospectivity

1st level	2nd level				
Gb + AGS + RS	Gb + AGS	Gb + RS	AGS + RS	Morphom. + RS	Gb + Morphom
Boca de Jaruco	SW de Punta Rubalcava	SW de San Antonio de Río Blanco	E de El Fraile	San Miguel de Casanovias	E de Peñalver
Jibacoa del Norte	Loma El Palenque	SE de El Perú	El Rubro		S de Cotorro
E de Aguacate	El Conde	SE de Brisas de Mar			

3.5 Conclusions

A version of the cartography of sectors with gaso-petroleum interest, linked to the conventional oil and gas of the TSU Placetas, is offered, based on the presence of a complex of indicator anomalies. This complex considers the following attributes: low-amplitude local gravimetric maximums; minimums of the K/Th ratio and local maximums of U (Ra) in its periphery; local maximums of the residual relief and RS anomalies.

A version of the geological–structural cartography of the study region is also presented, based on the gravimetric data.

These results constitute an essential non-seismic complementary information necessary for the geological–geophysical–geochemical–morphometric integration of the territory.

References

Colectivo de Autores (2009) Expediente Único del Proyecto 6004, "Exploración en la Franja Norte Petrolera Cubana". Inédito. Archivo, Centro de Investigaciones del Petróleo (Ceinpet), La Habana, Cuba

Colectivo de Autores (2010) Mapa Geológico Digital de la República de Cuba a escala 1:100000. Inédito. Instituto de Geología y Paleontología, Servicio Geológico de Cuba, La Habana

Fairhead JD, Ahmed S, Williams SE (2009) Tilt-depth: a simple depth-estimation method using first order magnetic derivatives. Search and Discovery Article #40390 (Adapted from poster presentation at AAPG International Conference and Exhibition, Cape Town, South Africa, October 26–29, 2008)

Jiménez de la Fuente L (2017) Resultados de la Teledetección (TDT) para la búsqueda de sectores gasopetrolíferos perspectivos en la región de Guanabo-Seboruco. Apuntes metodológicos. Inédito. Ceinpet, La Habana, 13 p

Mondelo F, Sánchez Cruz R y otros (2011) Mapas geofísicos regionales de gravimetría, magnetometría, intensidad y espectrometría gamma de la República de Cuba, escalas 1:2,000,000 hasta 1:50,000. Inédito. IGP, La Habana, 278 p

Pardo Echarte ME, Rodríguez Morán O (2016) Unconventional methods for oil & gas exploration in Cuba. Springer Briefs in Earth System Sciences. https://doi.org/10.1007/978-3-319-28017-2

Pardo Echarte ME (2016) Generalización Petrofísica de la Provincia Matanzas. Inédito. Centro de Investigaciones del Petróleo, La Habana, 5 p

Price LC (1985) A critical overview of and proposed working model for hydrocarbon microseepage. US Department of the Interior Geological Survey. Open-File Report 85-271

Sánchez Cruz R, Mondelo F y otros (2015) Mapas Morfométricos de la República de Cuba para las escalas 1: 1,000,000–1:50,000 como apoyo a la Interpretación Geofísica. Memorias VI Convención Cubana de Ciencias de la Tierra, VIII Congreso Cubano de Geofísica. Fuente: http://www.cgiar-csi.org/data/srtm-90m-digital-elevation

Saunders DF, Burson KR, Thompson CK (1999) Model for hydrocarbon microseepage and related near-surface alterations. AAPG Bull 83(1):170–185

Schumacher D (1996) Hydrocarbon-induced alteration of soils and sediments. In: Schumacher D, Abrams MA (eds) Hydrocarbon migration and its near-surface expression. AAPG Memoir 66, pp 71–89

Chapter 4
Results of Some Non-seismic Exploration Methods in Different Gaso-Petroleum Regions of Western and Central Cuba

Manuel Enrique Pardo Echarte and Osvaldo Rodríguez Morán

Abstract It is well documented that most hydrocarbon accumulations have escapes or microseepage that are predominantly vertical, as well as that they can be detected and mapped using various non-conventional and non-seismic exploration methods. The surface expression of hydrocarbon microseepage can take a variety of forms, which determine the development of several detection methods, both direct (hydrocarbon gas geochemistry) and indirect (***Redox Complex***), as well as non-seismic geophysical methods (gravimetry, magnetics, electrics, and airborne gamma spectrometry), morphometry, and remote sensing. Benefits in the use of non-seismic and non-conventional exploration methods, integrated with geological data and seismic, result in a better evaluation of prospects and exploration risk. The source materials are as follows: the gravimetric (Bouguer reduction 2.3 t/m^3) and aeromagnetic (reduced to pole) maps at a scale of 1:50,000; airborne gamma-spectrometric maps (channels K, Th and U (Ra)) at a scale 1:100,000; the Digital Elevation Model (90 × 90 m) and; digital maps of hydrocarbon shows and oil wells of the Republic of Cuba at a scale of 1:250,000. The processing consisted in the regional-residual separation of the gravimetric and morphometric fields, the calculation of the first vertical derivative of the gravimetric and aeromagnetic fields and of the K/Th ratio. A mapping of sectors of oil–gas interest in western and central Cuba related to the conventional oil of the Placetas Tectonic-Stratigraphic Unit and the Jurassic level is based on the presence of a complex of indicator anomalies. It considers the following attributes: subtle local gravimetric maximums (in or near regional minimums); minimum of the K/Th ratio and local maximums of U (Ra) at its periphery, and; local maximums of residual relief.

Keywords Non-seismic methods of hydrocarbon exploration · Gravimetry · Aeromagnetics · Airborne gamma spectrometry · Morphometric methods

M. E. Pardo Echarte (✉) · O. Rodríguez Morán
Centro de Investigaciones del Petróleo (Ceinpet), Churruca, no. 481, e/Vía Blanca y Washington,
El Cerro CP 12000, La Habana, Cuba
e-mail: pardo@ceinpet.cupet.cu

O. Rodríguez Morán
e-mail: ormoran2016@gmail.com

© The Author(s), under exclusive license to Springer Nature Switzerland AG 2019
M. E. Pardo Echarte et al., *Non-seismic and Non-conventional Exploration Methods for Oil and Gas in Cuba*, SpringerBriefs in Earth System Sciences,
https://doi.org/10.1007/978-3-030-15824-8_4

83

4.1 Introduction

The seismic exploration is insurmountable to provide structural and stratigraphic information, as well as for the cartography and the obtaining of images of traps and reservoirs. However, in various geological situations, seismic data provide little or no information about whether a trap is loaded with hydrocarbons or not. In other cases, when the acquisition is difficult and extremely expensive, or the quality of the information is poor due to geology or unfavorable surface conditions, it is the non-seismic exploration methods and, in particular, the unconventional methods of exploration, the only ones that can provide information on subtle stratigraphic traps. Some of the traditional onshore non-seismic geophysical methods of exploration for oil and gas consider potential fields (gravimetry and aeromagnetics) and, more limitedly, airborne gamma spectrometry (AGS). According to Dobrin and Savit (1988), Garland (1989) and Gubins (1997), gravimetry allows the geological–structural cartography of large geological units, the search for local structures in the sedimentary cover and, also, the detailing of the main tectonic features and other alignments with which the energy resources are linked. According to the same authors, in aeromagnetics, the ability to map geological–structural features is reinforced by the possibility of detecting and mapping anomalies of little amplitude with which the most superficial processes of secondary magnetite mineralization are linked to the occurrences of oil and gas. Also on them, the decomposition of clays in the soils produced by the same factor, the light hydrocarbons microseepage, causes the minimum radiation and K/Th ratio observed, with local increases of U (Ra) in its periphery, for whose cartography the AGS is useful.

The use of other unconventional exploration methods such as morphometry and geophysical–geochemical techniques of indirect detection (***Redox Complex***) are based on the association of positive residual geomorphic anomalies and an anomalous complex of physical–chemical attributes, respectively, on oil and gas occurrences (Pardo Echarte and Rodríguez Morán 2016).

The territory of Cuba is privileged by its contrasting alpine geology, its tropical climate, which conditions the presence of residual soils or in situ weathering crusts and the presence of an eminently flat relief. Cuba has an aeromagnetic and airborne gamma-spectrometric survey at 1:50,000 scale throughout the national territory and a gravimetric survey, at the same scale, in 80% of it. Also, it has digital elevation models (DEMs) 90×90 m and 30×30 m, so that adequate information coverage is available for the study of the regions to be investigated. The geological task posed to geological–geophysical processing and interpretation consisted in the mapping of possible new gaso-petroleum targets for the foundation of oil exploration in different onshore regions and blocks (east of Motembo [Block 9] and surroundings of Jarahueca Oilfield [Block 13]) from the west and center of Cuba. The mapping of the possible new gaso-petroleum targets is made from the presence of a complex of indicator anomalies (fundamentally, gravimetric, airborne gamma spectrometric, and morphometric). To this end, the gravimetric and aeromagnetic fields at 1:50,000 scale, the airborne gamma spectrometry at 1:100,000 scale of the different territories and the DEM (90 × 90 m) were processed. For reasons of confidentiality, maps without coordinates are presented.

4.1.1 Geological Premises that Support the Application of Non-conventional and Non-seismic Geophysical–Geochemical–Morphometric Methods of Exploration

From the point of view of Surface Geochemistry, according to Price (1985), Schumacher (1996), Saunders et al. (1999) and Pardo Echarte and Rodríguez Morán (2016), the geological premises that base the application of the non-conventional geophysical–geochemical–morphometric exploration methods are the following:

- The "Reducing Chimneys" are columns of mineralized rocks above the hydrocarbon deposits which were modified by the vertical migration of these and/or by some other association of reduced species (metal ions) which "oxidize," by microbial action, to create a reducing environment.
- The main products of the microbial oxidation of hydrocarbons (CO_2) and the microbial reduction of sulfur (H_2S) drastically change the pH/Eh of the system.

 Changes in pH/Eh result in changes in mineral stability:

- Precipitation of various carbonates.
- Decomposition of clays (as a consequence, increase the concentrations of silica and alumina).
- Precipitation of magnetite/maghemite, of iron sulfides (such as pyrrhotite and greigite) or coprecipitation of iron and/or manganese with calcite in carbonated cements on hydrocarbon deposits.

 The morphometric, geophysical, and geochemical response to previous mineral stability changes is as follows:

- Secondary calcium carbonate mineralization and silicification result in denser and erosion-resistant surface materials (formation of positive residual geomorphic and resistivity maximum anomalies).
- The decomposition of clay is responsible for the minimum radiation reported on the oil deposits: potassium is leached from the system toward the edges of the vertical projection of the hydrocarbon deposit, where it precipitates resulting in a "halo" of high values. Thorio remains relatively fixed in its original distribution within insoluble heavy minerals; hence, minimum K/Th ratios are observed, surrounded by peaks, on oil and gas deposits. In the periphery of the occurrences, maximums of U (Ra) are observed.
- The conversion of non-magnetic iron minerals (oxides and sulfides) into more stable magnetic varieties results in an increase in magnetic susceptibility, correlated

with the minimum of the Redox Potential, which justifies the integration of both techniques. Induced polarization anomalies are also observed.

- The arrival to the surface of the metallic ions contained in the hydrocarbons (V, Ni, Fe, Pb, and Zn, among others) condition the presence of a subtle anomaly of these elements in the soil and a slight change in coloration of the same which is reflected by anomalies of the spectral reflectance, facts that justify the integration of these techniques.

4.2 Materials and Methods

4.2.1 Information and Its Sources

The materials used and their sources are the following:

- Meshes of the gravimetric and aeromagnetic fields at 1:50,000 scale and, AGS (channels: It, U, Th, and K) at 1:100,000 scale of the Republic of Cuba (Mondelo et al. 2011).
- DEM (90 × 90 m) from the Republic of Cuba (Sánchez Cruz et al. 2015).
- Digital Maps of hydrocarbon shows and oil wells of the Republic of Cuba at a scale of 1:250,000 (Colectivo de Autores 2008, 2009, respectively).

4.2.2 Processing and Interpretation of Information

The processing of the geophysical-morphometric information was carried out using the software Oasis Montaj version 7.01.

The gravimetric field (Bouguer reduction, 2.3 t/m^3) was subjected to the regional-residual separation from the Upward Analytical Continuation (UAC) for heights of 500 and 2000 m (given by the order of depth of the possible oil and gas targets) and the first vertical derivative (VD). The search targets are characterized by subtle local maximums within or on the periphery of the regional minimums that correspond to the North Cuban Thrust Belt.

The aeromagnetic field underwent the reduction to pole and the VD, seeking to map different tectonic alignments.

For the AGS, the K/Th ratio was determined, indicating the minimums presumably linked to active zones of light hydrocarbons microseepage, and, indicating the local maximums of U (Ra) in its periphery.

The DEM (90 × 90 m) underwent a regional-residual separation from the UAC at 500 m (according to the experience of the main author in this type of work) to contour the zones of residual maximums, linked to the light carbonization and subsurface silicification processes that take place as a result of the active light hydrocarbons microseepage on oil and gas occurrences.

4.3   Results and Discussion

4.3.1   Region East of Motembo (Motembo-Carbonates, Block 9)

The geological model for the study area is conceived as the local uplift of a scale of the Placetas Tectonic-Stratigraphic Unit (TSU), represented by the Veloz Group, outcropping in the locality, which restricts the gaso-petroleum prospectivity of the sector to the oil of Jurassic level with an anhydrite seal (also Jurassic).

The establishment of gaso-petroleum interest of this sector is based on the presence of a complex of indicator anomalies that considers the following attributes:

- A subtle local gravimetric maximum in the proximity of a regional minimum (thick blue trace), which must be reflecting a positive local structure (due to the lifting of the denser carbonates);
- A potassium and ratio K/Th minimums (red trace) with a local maximum of U (Ra) (pink trace) in its periphery and;
- A local maximum of residual relief (black trace).

As complementary geological information, the presence of oil wells (black dots) in the region was considered (Colectivo de Autores 2009). All these attributes have been represented in the complex interpretation results of non-seismic geophysical-morphometric methods of Fig. 4.1. They reveal a sector of gaso-petroleum interest which is recommended to be validated by means of a recognition profile of the *Redox Complex*. In the same figure, to the SW of this locality, another sector of interest (Motembo SE), has been mapped, within the Motembo ultrabasic massif, which has been validated, positively, by a profile of the *Redox Complex* (blue line with NO-SE direction) for shallow naphtha in serpentines, Motembo type. The investigations carried out in this target are described, in detail, in another publication (Pardo Echarte and Cobiella Reguera 2017).

4.3.2   Region of the Surroundings of Jarahueca Oilfield (Block 13)

The geological model for the study area, according to Pérez Martínez et al. (2013), conceives the traps generated during the thrusts and with the transpressive redesign, considering the prospectivity for the conventional oil of the TSU Placetas. The main seal is considered the orogenic argillaceous sequences, and subordinately, with less relevance, the serpentinites that could constitute local seals.

For the establishment of sectors of gaso-petroleum interest in this region, it has been considered essential to characterize the anomalous complex of the Jarahueca Oilfield (A), from the presence of a set of indicator anomalies that consider the following attributes:

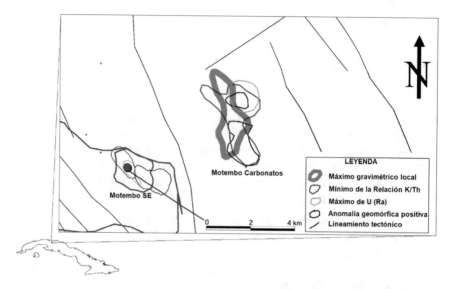

Fig. 4.1 Complex interpretation results of non-seismic geophysical–morphometric methods in the region east of Motembo (Motembo-Carbonates, Block 9)

- A local gravimetric maximum (thick black trace), which should be reflecting a positive local structure within the Jarahueca tectonic window (lifting of the denser ophiolites), coinciding with a magnetic anomaly which characterizes the outcrop ophiolites in that locality;
- Minimum of the K/Th ratio (thick blue trace) with local maximums of U (Ra) (red trace) at its periphery and;
- A local maximum of residual relief (black trace).

As complementary geological information, the presence of hydrocarbon shows (black pentagons) in the region was considered (Colectivo de Autores 2008). All these attributes have been represented in the complex interpretation results of the non-seismic geophysical-morphometric methods of Fig. 4.2. Although the anomalous complex described in A is very specific and, apparently, unrepeatable in the region, it finds a partial analogy in B anomalous complex (with the same structural position). Complexes C and D, different in their structural position, are characterized by significant anomalies of the K/Th ratio. The last three mentioned anomalous complexes (B, C and D) reveal, together with A, an extended zone of possible gaso-petroleum interest, which is recommended to be validated by recognition profiles of the *Redox Complex*.

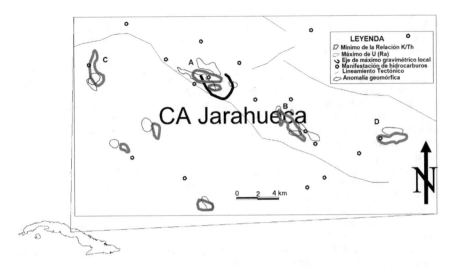

Fig. 4.2 Complex interpretation results of non-seismic geophysical-morphometric methods in the region surrounding the Jarahueca Oilfield (Block 13)

4.4   Conclusions

A cartography version of sectors of gaso-petroleum interest is offered in western and central Cuba. They are linked to the conventional oil of the Placetas TSU and to the Jurassic level and were established from the presence of a complex of indicator anomalies. It considers the following attributes: subtle local gravimetric maximums (in the proximity or within the regional minimums); minimums of the ratio K/Th and local maximums of U (Ra) in its periphery, and; local maximums of the residual relief.

References

Colectivo de Autores (2008) Mapa Digital de las Manifestaciones de Hidrocarburos de la República de Cuba a escala 1:250,000. Inédito. Centro de Investigaciones del Petróleo, La Habana

Colectivo de Autores (2009) Mapa Digital de los Pozos Petroleros de la República de Cuba a escala 1:25,0000. Inédito. Centro de Investigaciones del Petróleo, La Habana

Dobrin MB, Savit OH (1988) Introduction to geophysical prospecting, 4th ed. McGraw Hill International Editions, 867 pp

Garland GD (1989) Proceedings of exploration 87. In: Third decennial international conference on geophysical and geochemical exploration for minerals and groundwater, Special Volume 3. Ontario Geological Survey, 914 pp

Gubins AG (1997) Proceedings of Exploration 97. Fourth decennial international conference on mineral exploration. Prospectors and Developers Association of Canada, 1065 pp

Mondelo F, Sánchez Cruz R y otros (2011) Mapas geofísicos regionales de gravimetría, magnetometría, intensidad y espectrometría gamma de la República de Cuba, escalas 1:2,000,000 hasta 1:50,000. Inédito. IGP, La Habana, 278 p

Pardo Echarte ME, Cobiella Reguera JL (2017) Oil and gas exploration in Cuba: geological-structural cartography using potential fields and airborne gamma spectrometry. Springer Briefs in Earth System Sciences. https://doi.org/10.1007/978-3-319-56744-0

Pardo Echarte ME, Rodríguez Morán O (2016) Unconventional methods for oil & gas exploration in Cuba. Springer Briefs in Earth System Sciences. https://doi.org/10.1007/978-3-319-28017-2

Pérez Martínez Y, Valdivia CM y otros (2013) Proyecto I + D 7054, Etapa 1.4 "Informe final sobre fundamentación de pozo en el Bloque 13". Inédito Centro de Investigaciones del Petróleo, La Habana, 75 p

Price LC (1985) A critical overview of and proposed working model for hydrocarbon microseepage. US Department of the Interior Geological Survey. Open-File Report 85-271

Sánchez Cruz R, Mondelo F y otros (2015) Mapas Morfométricos de la República de Cuba para las Escalas 1:1,000,000–1:50,000 como apoyo a la Interpretación Geofísica. Memorias VI Convención Cubana de Ciencias de la Tierra, VIII Congreso Cubano de Geofísica. http://www.cgiar-csi.org/data/srtm-90m-digital-elevation

Saunders DF, Burson KR, Thompson CK (1999) Model for hydrocarbon microseepage and related near-surface alterations. AAPG Bull 83(1):170–185

Schumacher D (1996) Hydrocarbon-induced alteration of soils and sediments. In: Schumacher D, Abrams MA (eds) Hydrocarbon migration and its near-surface expression. AAPG Memoir 66, pp 71–89

Chapter 5
Morphotectonic Regionalization in the Seas of South Cuba from the Digital Elevation Model 90 × 90 m

Manuel Enrique Pardo Echarte

Abstract The scenario of the study area includes the seas to the south of Cuba: shallow, transitional and deep waters, characterized by the Batabanó, Ana María and Guacanayabo gulfs, parts of the Yucatan Basin, the Cayman Ridge and the Cayman Trench (on the edge of the Bartlett Fault). For this territory, with a low degree of seismic study, it was very important to count with other complementary non-seismic exploratory tools, like Digital Elevation Model, that allowed a preliminary assessment of the prospective areas. Then, the geological task posed to the morphometric processing and interpretation consists of the morphotectonic regionalization, with precision of the morpho-structural evidences of the so-called Camagüey Trench existence and, the establishment of possible sectors of oil and gas interest linked to the presence of geomorphic anomalies, presumably Indicators. For such purposes, the Digital Elevation Model 90 × 90 m was processed. As a result, the Morphotectonic Regionalization of the South Cuba marine territory was carried out, recognizing three types of regions: shallow waters (less than 100 m), transitional waters (greater than 100 and less than 3000 m), and deep waters (greater than 3000 m), from which could be characterized the different physiographic features already known. According to the interpretation, the existence of obvious signs of a trench and, therefore, of a paleo subduction zone in southern Cuba, is not recognized. Taking into account the range of amplitude of the geomorphic anomalies, if there is an active petroleum system, the areas of greatest prospective interest would correspond to the three gulfs (shallow waters) and, to the southwest and south of Batabanó Gulf and the northern limit of Cayman Ridge (transitional waters). The anomalies in deep water are vetoed by the huge water strain.

Keywords South of Cuba · Offshore hydrocarbon exploration · Morphotectonic regionalization · Digital Elevation Model · Geomorphic anomalies

M. E. Pardo Echarte (✉)
Centro de Investigaciones del Petróleo (Ceinpet), Churruca, no. 481, e/Vía Blanca y Washington, El Cerro CP 12000, La Habana, Cuba
e-mail: pardo@ceinpet.cupet.cu

© The Author(s), under exclusive license to Springer Nature Switzerland AG 2019   91
M. E. Pardo Echarte et al., *Non-seismic and Non-conventional Exploration Methods for Oil and Gas in Cuba*, SpringerBriefs in Earth System Sciences, https://doi.org/10.1007/978-3-030-15824-8_5

5.1 Introduction

5.1.1 Geographic Location and General Characteristics

The satellite scenario of the study area (Fig. 5.1) includes the seas south of Cuba: shallow, transitional, and deep waters, characterized by the Batabanó, Ana María and Guacanayabo gulfs, parts of the Yucatan Basin, the Cayman Ridge and of the Cayman Trench (in the limit of the Bartlett Fault).

This region belongs to the so-called Southern Cuban Petroleum Province (SCPP) where, according to Morales Carrillo et al. (2011), several types of basins are present:

- Land-Shallow Water (Vertientes-Ana María and Cauto-Guacanayabo)
- Shallow waters (Los Canarreos)
- Transitional and Deep Waters (Prolongation of Cauto-Guacanayabo and Yucatan, respectively).

The main physiographic features are the Yucatan Basin and the Cayman Ridge. The Yucatan Basin in its western central part is a turbiditic plain, while the eastern and southern parts are less deep and of varied relief. Its western and northern limits, against Belize and Mexico, and against Cuba are stepped and widely faulted. For its part, the Cayman Ridge, according to some authors, could be a fragment of the Nicaraguan Rise, separated from it due to the opening of the Cayman Trench, or an elevated block composed mainly of carbonates from shallow waters, according to others.

According to Morales Carrillo et al. (2011), the SCPP proposes a type of source rock of age and compositions different from those already known in the Northern Province, among these elements are:

Fig. 5.1 Satellite image with the scenario of the study region. With a blue arrow, we can see the so-called Camaguey Trench by Pardo (2009)

- Gas shows in reef limestones in Cauto-Guacanayabo.
- Gas shows in the Canarreos Archipelago.
- It has been established by biomarkers the presence of organic matter originating in higher plants associated with source rocks in the samples taken from several cays located in Jardines de la Reina Archipelago, whose age can be Upper Cretaceous and even younger.

In the same way, according to López Quintero (2010), during the detection of oil shows on the sea surface at Ana María and Guacanayabo gulfs, in the last decade of the last century, the presence of oil on the shores of beaches, in the form of bitumen and tar balls, was established. The results of the study of the samples taken and the correlation with known oils of the three established Cuban families showed that the bitumen found on the beaches of the Tortuga Shoals, Rabihorcado, Caguma, and Paloma cays have a totally different origin than the known families in Cuba.

Finally, Morales Carrillo et al. (2011) conclude that the active petroleum systems that are forecast in this region should have given rise to oil and gas deposits, the latter associated with biogenic gas and methane hydrates.

On the other hand, according to Pardo (2009), in the south of Cuba, there is a sublatitudinal trench called "Camagüey Trench," located to the north of the Cayman-Bartlett system (Fig. 5.1), which the author considers is not more than the trace of an old subduction zone on which an accretion prism developed. As it is known, the sediments located in the accretion prisms when entering the lithosphere can give rise to source rocks that generate hydrocarbons if have the amount of organic matter required for this. Therefore, Morales Carrillo et al. (2011) states that, taking into account what Pardo (2009) expressed, as well as the presence of gas and hydrocarbon shows that coincide spatially with the zone of occurrence of the so-called Camagüey Trench, the presence of source rocks can be expected in the south of Cuba associated with this event.

For this territory, with a low degree of seismic study, it is very important to have other complementary non-seismic exploratory tools that allow a preliminary assessment of the prospective areas. This is the main objective of this work.

5.1.2   Geological Task

The geological task posed to the processing and morphometric interpretation of the study region consisted of morphotectonic regionalization, with precision of the morpho-structural evidences of the existence of the so-called Camagüey Trench and, in the establishment of possible sectors of gaso-petroleum interest linked to the presence of presumably indicating geomorphic anomalies. For such purposes, the Digital Elevation Model (DEM) 90 × 90 m was processed (Sánchez Cruz et al. 2015).

5.2   Theoretical Framework

5.2.1   Geological Premises

According to Pardo Echarte et al. (2018), the geological premises that base the application of the DEM and other non-conventional geophysical methods in the detection of possible microseepage of light hydrocarbons and their applications for offshore hydrocarbon exploration are the following:

- Microseepage of light hydrocarbons (vertical migration of these to the surface) encourages the formation of mineralized rock columns above the oil and gas deposits (Price 1985; Pardo Echarte and Rodríguez Morán 2016).
- The migrating hydrocarbons are oxidized by microbial action, creating a reducing environment (Reducing Chimneys) and producing, mainly, CO_2 and H_2S, which drastically changes the pH/Eh of the system.
- Changes in pH/Eh result in changes in mineral stability, causing secondary mineralization of calcium carbonate and silicification, which result in denser, erosion resistant, and resistive surface materials (formation of positive geomorphic anomalies (residual DEM) and maximum resistivity).
- These changes in pH/Eh also cause the precipitation of magnetite/maghemite, and of iron sulfides (pyrite, pyrrhotite and greigite) or coprecipitation of iron and/or manganese with calcite in carbonated cements on hydrocarbon deposits, which result in an increase in magnetic susceptibility and polarizability (producing magnetic and induced polarization anomalies).

The coincidence of all these anomalies, geomorphic and geophysical, allows us to argue the combined use of unconventional exploration methods complementary to seismic investigations. Thus, it is proposed to carry out complementary investigations for offshore hydrocarbon exploration integrated by a complex of: DEM (90 × 90 m), magnetic survey and induced polarization (IP). It should be noted that the geomorphic anomalies of interest (residual maximums) vary between 1 and 25 m; while the magnetic, polarizability, and resistivity anomalies must be clearly observable.

The use of the DEM (90 × 90 m) provides a quick way to preliminary offshore oil and gas exploration being useful to focus the areas to be raised with complementary non-conventional methods (Magnetic and IP) whose complex results would argue the volumes of the most expensive 2D-3D detail seismic.

5.3 Materials and Methods

5.3.1 Information, Its Sources, and Processing

The DEM (90 × 90 m) used in this work was taken from Sánchez Cruz et al. (2015) with a source at: http://www.cgiar-csi.org/data/srtm-90m-digital-elevation + ETOPO 2. It is presented in Fig. 5.2.

The processing of the morphometric information was carried out using the software Oasis Montaj version 7.01. This consisted of the regional-residual separation of the submarine relief allowing to emphasize anomalies of smaller vertical amplitude, from the calculation of the Upward Analytical Continuation to 500 m, subtracting this from the observed relief (DEM), according to the author's experience. So, there indicated the local maximums, linked to the possible processes of light carbonization and subsurface silicification that take place on the active microseepage of light hydrocarbons. The residual DEM at 500 m is presented in Fig. 5.3.

The morphotectonic alignments were drawn from the limits of the anomaly chains, linearity zones of the contours, with a marked horizontal gradient, as well as their abrupt bending.

The residual maximums with horizontal dimension greater than 1000 m, with little linear extension, were plotted and indicated with their amplitude.

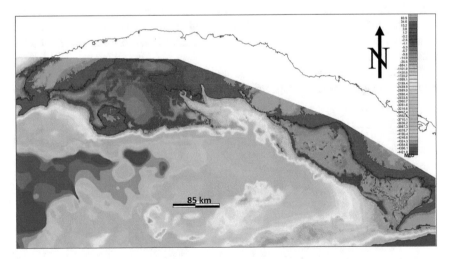

Fig. 5.2 Digital Elevation Model (90 × 90 m) of the marine territory of South Cuba, taken from Sánchez Cruz et al. (2015)

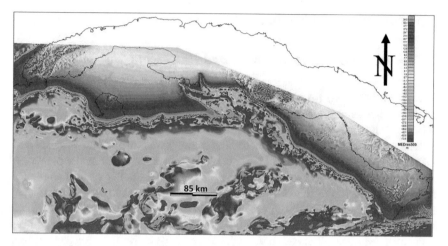

Fig. 5.3 Residual DEM at 500 m from the marine territory of South Cuba

5.4 Results

The morphotectonic regionalization from the observed submarine relief (DEM) (Fig. 5.4) allowed the establishment of three types of regions: shallow waters (less than 100 m), transitional waters (greater than 100 and less than 3000 m), and deep water (greater than 3000 m). From them, it could be characterized the areas of Batabanó, Ana María and Guacanayabo gulfs, Cayman Ridge, Yucatan Basin, and the Bartlett Fault.

Judging from the distribution of the morphotectonic alignments (Fig. 5.5), the zones of greater gradient (slope) follow the boundary that separates the shallow

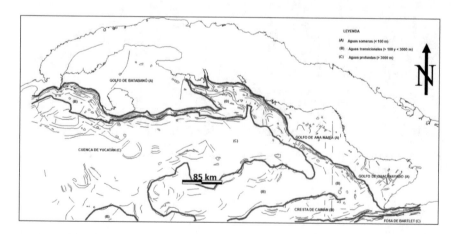

Fig. 5.4 Morphotectonic Regionalization in the seas south of Cuba

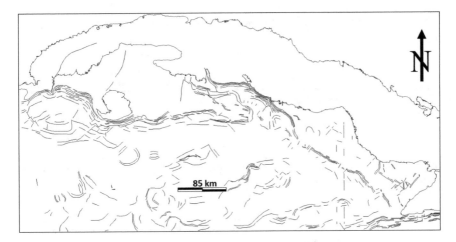

Fig. 5.5 Morphotectonic alignments in the seas south of Cuba

waters from the transitional ones all along the south of Cuba, which does not coincide with the position, practically linear (slightly curved), marked for the so-called Camagüey Trench by Pardo (2009) in Fig. 5.1. This allows us to state that, according to the interpretation of the morphometric data made the existence of evident signs (witnesses) of this trench in the south of Cuba and, therefore, of the supposed paleo subduction zone is not recognized.

The regularities of the Morphotectonic Regionalization of South Cuba, taking into account the distribution and amplitude of the positive residual geomorphic anomalies, have been synthesized in Table 5.1. Considering the anomalies range of amplitude, if there is an active petroleum system, the areas of greatest prospective interest would correspond to the three gulfs (shallow waters) and, to the southwest and south of the Batabanó Gulf and the Cayman Ridge northern limit (transitional waters). The anomalies corresponding to these five sectors are shown in Figs. 5.6, 5.7, 5.8, 5.9, and 5.10. The anomalies in deep waters, although of greater dimensions, are vetoed by the enormous water strain.

5.5 Conclusions

- The morphotectonic regionalization of South Cuba marine territory was carried out recognizing three types of regions: shallow waters (less than 100 m), transitional waters (greater than 100 and less than 3000 m), and deep waters (greater than 3000 m). From this, it was possible to characterize the areas of Batabanó, Ana María and Guacanayabo gulfs, Cayman Ridge, Yucatán Basin, and Bartlett Fault.

Table 5.1 Regularities of the South Cuba marine territory Morphotectonic Regionalization

Region		Predominant direction of anomalies	Predominant amplitude (range) of anomalies (m)	Predominant transversal dimension of anomalies (m)
Shallow waters (<100 m)	Batabanó Gulf	NO–SE (Cuban)	3–10	2600–3600
	Ana María Gulf	NO–SE (Cuban)	7–23	2000–3000
	Guacanayabo Gulf	SO–NE (transversal)	17	1500–2500
Transitional waters (100–3000 m)	SW and S of Batabanó Gulf	Following the slope	−5 a 27	1000–3000
	Cochinos Bay–Trinidad	Following the slope	13–85	1200–3500
	SW and S of Ana María Gulf	Following the slope	16–89	1500–3500
	SW of Guacanayabo Gulf	Following the slope	−19 a 79	1500–2500
	Cayman Ridge	Following north limit	3–18	1500–3500
Deep Waters (Yucatan Basin) (>3000 m)	Proximity to the north limit	Variable	−30 a 25	1500–4500
	Proximity to the south limit	Variable	0–34	1500–4500

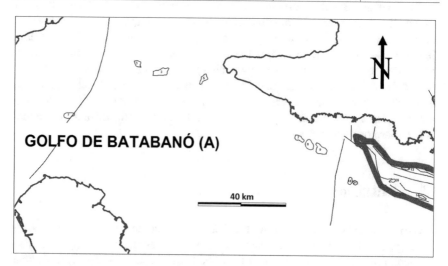

Fig. 5.6 Positive local geomorphic anomalies, with their amplitude, in the Batabanó Gulf

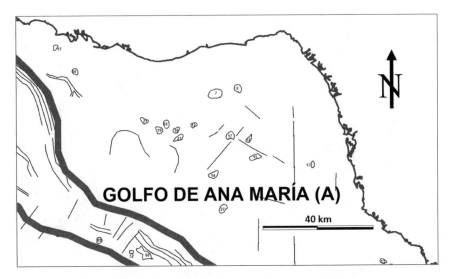

Fig. 5.7 Positive local geomorphic anomalies, with their amplitude, in the Ana María Gulf

Fig. 5.8 Positive local geomorphic anomalies, with their amplitude, in the Guacanayabo Gulf

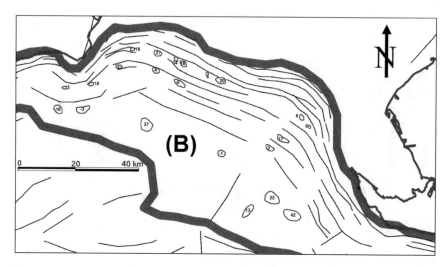

Fig. 5.9 Positive local geomorphic anomalies, with their amplitude, to the S–SW of Batabanó Gulf

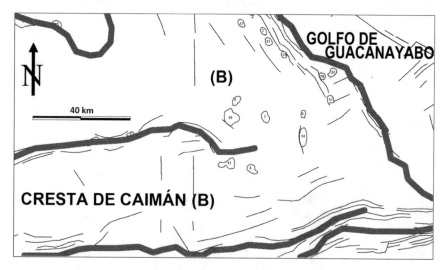

Fig. 5.10 Positive local geomorphic anomalies, with their amplitude, at the Cayman Ridge northern limit

- According to the interpretation of the morphometric data, the existence of obvious signs (witnesses) of a trench and, therefore, of a paleo subduction zone in southern Cuba is not recognized.
- Taking into account the anomalies range of amplitude, if there is an active petroleum system, the areas of greatest prospective interest would correspond to the three gulfs (shallow waters), to the S–SW of Batabanó Gulf and to the northern limit of Cayman Ridge (transitional waters). The anomalies in deep water are vetoed by the huge water strain.

References

ETOPO 2. Fuente: http://worldgrids.org/doku.php/wiki:dem_sources

López Quintero JO (2010) ¿Nueva familia de petróleos cubanos en la plataforma marina sur de Cuba? Una incógnita de la geología cubana. Memorias V Jornada Calidad del Centro de Investigaciones del Petróleo, Cuba, 5 pp

Morales Carrillo I, López Rivera JG y Delgado O (2011) La provincia sur petrolera cubana, un reto para la exploración de hidrocarburos. CUARTA CONVENCIÓN CUBANA DE CIENCIAS DE LA TIERRA, GEOCIENCIAS'2011. Memorias en CD-ROM, La Habana, 4 al 8 de abril de 2011. ISBN 978-959-7117-30-8, 9 pp

Pardo G (2009) The geology of Cuba. AAPG Studies in Geology #58. Published by The American Association of Petroleum Geologists, Tulsa, Oklahoma, 268 pp

Pardo Echarte ME, Rodríguez Morán O (2016) Unconventional methods for oil & gas exploration in Cuba. Springer Briefs in Earth System Sciences. https://doi.org/10.1007/978-3-319-28017-2

Pardo Echarte ME, Reyes Paredes O, Súarez Leyva V (2018) Offshore exploration of oil and gas in Cuba using Digital Elevation Models (DEMs). Springer Briefs in Earth System Sciences, 53 pp. ISBN 978-3-319-77154-0 ISBN 978-3-319-77155-7 (eBook). https://doi.org/10.1007/978-3-319-77155-7

Price LC (1985) A critical overview of and proposed working model for hydrocarbon microseepage. US Department of the Interior Geological Survey. Open-File Report pp 85–271

Sánchez Cruz R, Mondelo F y otros (2015) Mapas Morfométricos de la República de Cuba para las Escalas 1:1,000,000–50,00 como apoyo a la Interpretación Geofísica. Memorias VI Convención Cubana de Ciencias de la Tierra, VIII Congreso Cubano de Geofísica. http://www.cgiar-csi.org/data/srtm-90m-digital-elevation + ETOPO 2

Index

Printed in the United States
By Bookmasters